1 MONTH OF
FREE
READING

at

www.ForgottenBooks.com

By purchasing this book you are eligible for one month membership to ForgottenBooks.com, giving you unlimited access to our entire collection of over 1,000,000 titles via our web site and mobile apps.

To claim your free month visit:
www.forgottenbooks.com/free924783

ISBN 978-0-260-05448-7
PIBN 10924783

TABLES

FOR THE

Determination of Common Minerals

CHIEFLY BY THEIR

PHYSICAL PROPERTIES

WITH CONFIRMATORY CHEMICAL TESTS.

BY

W. O. CROSBY,

ASSISTANT PROFESSOR OF MINERALOGY AND LITHOLOGY IN THE

MASSACHUSETTS INSTITUTE OF TECHNOLOGY.

SECOND EDITION.

BOSTON:
J. ALLEN CROSBY,
1891.

PRINTING OFFICE OF THE PUBLISHER,
JAMAICA PLAIN, BOSTON, MASS.

PREFACE.

Although the published tables for the determination of minerals are very numerous, yet an experience of five years in the Mineralogical Department of the Massachusetts Institute of Technology and in the Teachers' School of Science of the Boston Society of Natural History has convinced me that several important features are wanting in the best tables, for the use of beginners and general students, with which I am acquainted. The primary objects of any system of de-derminative mineralogy should be : (1) to enable the student to identify certainly and easily such minerals as he is likely to meet; and (2) to cultivate his powers of observation and discrimination, and increase his familiarity with the various species and natural groups of minerals. The first of these primary objects has been, rightly enough, the chief aim of previous tables; but it is to be hoped that, as regards its direct educational value, determinative mineralogy will not always be, as it is now, far behind determinative or analytical botany.

The principal defects of the best tables now in use, so far, at least, as the general student is concerned, are that they have too wide a scope, embracing all or a very large proportion of the known species of minerals, and that the determinations depend almost entirely upon the chemical behavior of the species.

It is scarcely possible to overstate the enormous disparity, as regards their abundance, between the two or three hundred common minerals and the one thousand or more rare species. The former are found abundantly in many localities, and in all good collections, and are, generally speaking, the only species the student will meet or have occasion to identify; while the latter are very restricted in distribution, being often known from only a single locality, occuring for the most part in impure or very minute specimens, and being inadequately represented even in the largest and most complete collections.

The advantages of limiting determinative tables to the common species are that they are then far less voluminous and complicated, the determinations can be made more readily and accurately, and the use of difficult or uncertain tests is avoided. The only disadvantage is that there must often remain the possibility that the specimen in hand belongs to one of the rare species not included in the tables. But, as already explained, this difficulty is very small; and it can not be entirely obviated by the use of the most perfect tables, since it is probable that a large proportion, perhaps a majority, of the rare species are still unknown and unnamed.

Chemical or blowpipe tests are valuable, since the indications which they afford are usually very definite and precise ; and since they direct the student's attention particularly to the chemical composition and behavior of the species he is investigating, which is of great importance. Nevertheless, there are several good reasons why they should occupy a subordinate place in a system of determinative mineralogy intended for general use. Not the least of these is their inconvenience, since they require, in the aggregate, a considerable amount of apparatus. and re-agents, in other words, a blowpipe laboratory. This consideration, and the knowledge of chemistry which the system demands, practically restricts the instruction to high schools and colleges ; and even then the chemical system is not found to be practicable in after life, when the student of mineralogy no longer has access to the facilities. afforded by the school. Every teacher of blowpipe analysis must have noted and lamented the fact that, as a rule, the elaborate system so carefully taught is used, after the students leave school, only by the few who become professional mineralogists, chemists or miners.

Perhaps the most serious objection to the chemical tables is that their use has comparatively little tendency to render their aid unnecessary, by making the student more familiar with the external appearances of minerals. Unquestionably, determinative tables are, at the best, a necessary evil. It were far better to recognize minerals on sight, by their structural and physical characters ; and, other things being equal, the preference should be given to that system of determination which promises the largest development of this power in the student.

These tables are, then, an attempt to determine about two hundred minerals by their more obvious physical and structural features, with confirmatory chemical tests. The latter will not usually be required when the specimens are pure and well characterized. But they are, as a rule, so simple and decisive that their use is strongly recommended whenever convenient and the determination is not otherwise perfectly satisfactory. Only those tests have been selected requiring the minimum of apparatus, re-agents and previous chemical training, with a view to adapting the tables to the use of common schools and private students.

The various properties of minerals, and all the chemical and blowpipe tests referred to in the tables, are fully explained in the introduction, which also includes an outline of the classification of minerals ; so that, although this little work is not in any sense a manual of mineralogy, constant reference to the more comprehensive treatises is avoided.

Boston Society of Natural History,
 Boston, November 1, 1886.

INTRODUCTION.

The properties or natural characteristics of minerals, by which they are known and recognized, may be arranged in three classes, as follows :—

1. Properties relating to the form and structure (crystallization, etc.) of minerals—*morphologic properties.*

2. Properties relating to the action on minerals of the various physical forces—*physical properties.*

3. Properties relating to the composition and chemical behavior of minerals—*chemical properties.*

For the sake of convenience, and to indicate their subordinate position in these tables, the chemical properties are described last, although in a systematic treatise on mineralogy this class would properly come first.

MORPHOLOGIC PROPERTIES OR CRYSTALLOGRAPHY.

The principal forms and structures of minerals are those due to their crystallization; but there are other kinds that are quite independent of crystallization. Hence crystallography is only a part, although the main part, of the general science of the forms and structures or morphology of minerals.

Definition of a Crystal.—A crystal is a natural solid bounded by plane surfaces symmetrically arranged with reference to certain imaginary lines or directions of growth passing through its centre and called the axes. Most crystals break or cleave with great ease in certain definite directions; and this geometric splitting, or crystalline cleavage, as well as certain optical properties of crystals, proves that crystallization means regularity of internal structure as well as of external form.

Degrees of Individualization of Crystals.—Crystals are the mineral individuals, and several degrees of distinctness and perfection of crystallization or individualization are recognized, as follows : When the crystals are distinct or separate and so nearly perfect that their proper forms may be clearly recognized, the mineral is *crystallized.* When we have a confused mass showing crystal-faces or planes and cleavage planes, but no perfect crystals (rock-salt and white marble), it is *crystalline* or *massive.* When crystalline form and cleavage are both entirely wanting to the unaided eye, but the specimen exhibits the phenomenon of double refraction when a thin section is viewed by polarized light (chalcedony), it is *cryptocrystalline* or *compact.* When mineral matter is entirely devoid of crystallization (opal and obsidian), it is described as *amorphous* or wholly unindividualized.

Forms of Crystals.—This is the most important and at the same time the most difficult section of crystallography; and the student is particularly recommended to refer to a standard treatise, such as Dana's Text-book of Mineralogy. But little more can be attempted here than an explanation of the terms used in the tables. The numerous forms of crystals are arranged in six systems, based on the relative lengths, positions and numbers of the axes. Each system has a number as well as a name, and in the tables they are indicated by the Roman numerals.

I. Isometric System.—The simplest form in this system is the *cube*, with six square sides or planes. Then comes the *octahedron* with eight triangular planes, the *dodecahedron* with twelve rhombic planes, the *trigonal trisoctahedron* with twenty-four triangular (trigonal) planes, the *tetragonal trisoctahedron* with twenty-four quadrangular (tetragonal) planes, the *tetrahexahedron* with twenty-four triangular planes arranged on the plan of the cube instead of the octahedron, and lastly the *hexoctahedron* with forty-eight planes. All of these forms, except the cube and dodecahedron, may occur with only half the planes developed, giving what are called *hemihedral* (half-faced) forms, which may have distinct names, the hemi-octahedron being also called the *tetrahedron*, etc. In general, both the number and arrangement of the planes are quite clearly expressed in the crystallographic names.

II. Tetragonal System.—The *square* or *tetragonal prism* is the simplest or fundamental form of this system. There is also an eight-sided or *ditetragonal prism*. The end planes of the prisms are called the *basal planes*. The remaining forms are the *square* or *tetragonal pyramid*, resembling the octahedron of the isometric system lengthened or shortened in the direction of one axis, and the *ditetragonal* or eight sided *pyramid*. The crystallographic pyramid is double, being equivalent to two geometric pyramids placed together, base to base; and it contains, in each case, twice as many planes as the name indicates.

III. Hexagonal System.—In this system, as the name implies, the fundamental form is the *hexagonal prism*. The holohedral forms are strictly analogous to those of the tetragonal system. Thus we have the *hexagonal* and *dihexagonal prisms*, *basal planes*, and *hexagonal* and *dihexagonal pyramids*. The most important of the hemihedral forms are the *hemihexagonal pyramid* or *rhombohedron*, a form bounded by six rhombic planes and resembling an oblique cube; and the *hemidihexagonal pyramid* or *scalenohedron*.

IV. Orthorhombic System.—The fundamental form here is the *ortho* (right) *rhombic prism*, which gives the name to the system. Then comes the right rectangular prism, consisting of two narrow planes called *brachypinacoids* and two broad planes called *macropinacoids*. The two end planes, as before, are the *basal planes*. The eight inclined or sloping planes corresponding in position to the orthorhombic prism planes make the *orthorhombic pyramid*; while the four inclined planes corresponding to the brachy and macro pinacoids, respectively, are the *brachy* and *macro domes.*

V. Monoclinic System.—This system is essentially similar to the last, except that one of the axial intersections is oblique, so that the forms all incline in one direction (monoclinic). The prefixes of the names pinacoid and dome are *ortho* and *clino,* instead of macro and brachy; and we call the prism and pyramid *monoclinic,* instead of orthorhombic.

VI. Triclinic System.—The forms of this system are analogous to those of the fourth and fifth systems; but are easily distinguished by the fact that the angles are all oblique.

When all the planes on a crystal are of one kind, it is described as a simple form and given one crystallographic name. But when they are of two or more kinds it is called a compound form, and requires for its accurate description as many crystallographic names as there are kinds of planes. Thus, on the ordinary crystal of quartz there are two kinds of planes, and we describe it as a combination of the hexagonal prism and the hexagonal pyramid.

These compound *forms* must be carefully distinguished from the compound or (*twin*) *crystals.* The quartz crystal, although described as a compound form, because there are two forms or kinds of planes, is a single, simple crystal. But if two such crystals should grow together in a regular manner, the result would be a double or twin crystal.

Structures of Crystals.—The only topic that need be noticed under this head is the regular splitting exhibited by most crystals—

Crystalline Cleavage, which is scarcely less important to the student of determinative mineralogy than the external forms of crystals.

Amorphous bodies, such as glass and opal, being essentially homogeneous, possess no planes of least cohesion, but are equally strong in all directions. Consequently, when broken they yield fragments of very irregular forms. But the regular molecular structure of crystals, as already stated, is usually indicated by definite directions of easy splitting or cleavage; so that the fragments present on one or more sides flat lustrous surfaces. Cleavage is independent of hardness, being almost equally perfect in diamond and talc, the hardest and softest of minerals. The cleavage directions or planes are always parrallel to some actually occuring, or to some possible, external planes of the crystal; and are usually limited to the simpler forms in each system. Consequently it is often possible to determine all the more important features of the crystallization of a mineral from the examination of small cleavage fragments.

There are different degrees of cleavage. When the mineral splits very easily, yielding smooth and · brilliant surfaces, like mica, the cleavage is *perfect* or *eminent*; and the inferior degrees are described as *distinct, indistinct* or *imperfect, interrupted, in traces, difficult, etc.*

The different cleavage directions are named after the external planes of the crystals with which they are parallel. Thus, in the isometric system we have *cubic, octahedral* and *dodecahedral* cleavages. The principal kinds of cleavage in the other systems are the *prismatic, basal* and *pyramidal,* besides the *rhombohedral* cleavage of the third system and the *pinacoidal* cleavages of the fourth, fifth and sixth systems.

Forms and Structures of Mineral Aggregates.—Minerals frequently occur in masses having a more or less definite external form and internal structure, the form and structure of which are not due wholly, if at all, to crystallization, *i. e.,* to a regular geometric arrangement of the mineral molecules, but are the product, in part at least, of forces different from those by which crystals are made. These masses or aggregates are usually uncrystalline, and are never single crystals or regular compounds of crystals.

External Forms of Mineral Masses.—When mineral matter is deposited in parallel layers on uneven surfaces, the upper surface of the deposit, like snow that has fallen on uneven ground, presents smoothly rounded hummocks or elevations. If these are small and somewhat grape-like in outline, the form is described as *botryoidal.* When the rounded prominences are larger, the form is called *mammillary.*

Closely related to these deposition forms are those produced when water holding mineral matter in solution falls, drop by drop, from an overhanging surface of rock. The pendant cone or column thus produced is called a *stalactite,* and the form is described as *stalactitic.* The low mound or layer made by the water dripping from the point of a stalactite upon the rocky floor below is called a *stalagmite,* the adjective term being *stalagmitic.* The porous deposits formed when reeds, grass, moss, and other kinds of vegetation are encrusted by mineral solutions is called a *tufa,* and its form and structure are described as *tufaceous.*

A round mass or nodule produced by the segregation of mineral matter in the body of a rock is called a *concretion,* the form being *concretionary.* If the concretions are small, about like peas, the form is called *pisolitic;* but if they are as small as mustard seed it is *oolitic.* Hollow concretions are called *geodes.* The *amygdaloidal* (almond-like) form results from the deposition of mineral matter in the vesicles or steam-holes of lava. In the stains of iron and manganese oxides known as dendrite, in moss agate, often in native copper, and on a frosted window pane, we have forms which are called *dendritic*, *arborescent* and *mossy*. These are usually more or less dependent upon the crystallization; and from these we pass easily to the *reticulated* and *plumose* forms. The native metals, especially, occur in wire-like or thread-like shapes, and are described as *filiform* or *capillary*; while slender, needle-like crystals are *acicular*. When a surface is thickly set with crystals of uniform size, they are described as *implanted crystals*; and when the implanted crystals are very small, the form is called *drusy*.

Internal Structures of Mineral Masses.—We have, first, the different kinds of *granular* structure. The grains are usually merely small, imperfect crystals, and there is a perfect gradation from the most *coarsely* to the most *finely* granular kinds. When the grains become invisible to the naked eye, the structure is *compact* or *impalpable*; and when no trace of a granular structure can be detected, even with the microscope, the mineral is *glassy* or *vitreous*. In this state it may be crystalline, like vitreous quartz, or amorphous, like obsidian.

The *lamellar* structures come next in importance. We properly distinguish at the outset those masses in which the lamination is entirely independent of crystallization, from those in which it is not. In the first class this structure is commonly described as *banded*, and in the second as *foliated*. The *banded* structure occurs commonly with the botryoidal, stalactitic, stalagmitic and geoditic forms, and the layers may be straight (plane) or curved and concentric. The *foliated* structure, on the other hand, is simply an easy splitting in parallel planes due to very perfect cleavage in one direction. The cleavage is usually basal; and foliation has its best development in the micas and allied minerals.

The *fibrous* structure is the third principal kind. It also depends upon the crystallization; for, as we may usually regard the granular masses as aggregates of short, thick crystals; the foliated masses as aggregates of flat, tabular crystals; so the fibrous masses may be regarded as examples of very slender, attenuated crystallization—prismatic crystallization carried to an extreme. When the fibres are large and distinct the structure is called *columnar* or *bladed*.

In the tables, not only the system in which the mineral crystallizes, but any marked habit of crystallization, and it is more characteristic uncrystalline or massive forms are briefly indicated.

The habit of crystallization is called *distinctly* prismatic or *distinctly* tabular etc., only when the development of these forms is very marked, the prisms, for example, being several or many times longer than thick.

PHYSICAL PROPERTIES.

The topics to be considered here are those characteristics of minerals depending upon their relations to the physical forces, such as cohesion, elasticity, light, heat, electricity, etc.

Properties relating to Cohesion and Elasticity.—*Cohesion* is the resistance which any body offers to a force tending to separate its molecules or particles, either by breaking or abrasion. The principal properties relating to this force are *cleavage, fracture,* and *hardness.*

Elasticity, on the other hand, is the force which tends to bring the molecules of a body back into their original positions when they have been disturbed. Upon elasticity depends, for the most part, the different degrees of tenacity possessed by minerals.

Cleavage and Fracture.—When minerals are broken, if the divisions are determined in direction and character by the crystallization, yielding smooth, plane surfaces parallel to the external planes of the crystal, the breaking is called *cleavage*; but if the divisions are independent of the crystalline form, and usually more or less irregular, the breaking is called *fracture.*

Although cleavage planes are directions in crystals along which the molecules separate readily, and cleavage is, in this respect, evidently related to cohesion, yet it is far more important as a manifestation of the molecular structure of crystals; and its characteristics have, therefore, been described in the preceding section (see page 5). When the cleavage is perfect, true fracture is difficult to obtain, on account of the strong tendency of the breaking to follow the directions of least resistance, *i. e.*, the cleavage directions. The most important kinds of fracture are: the *conchoidal,* the mineral breaking with curving concavities resembling the valve of a shell; the *even,* when the surface of the fracture is approximately smooth; the *uneven*; the *earthy,* breaking like clay or chalk; and the *hackly, splintery,* etc. In the tables, the fracture is given only where the cleavage is very imperfect or entirely wanting.

Hardness.—The hardness of a mineral is the resistance which it offers to abrasion. Hardness, however, is a purely relative term, and hence mineralogists have found it necessary to select certain minerals to be used as a standard of comparison for all others. This scale of hardness consists of ten minerals showing a regular gradation in hardness from talc, which is one of the softest of minerals, to diamond, the hardest of all minerals, as follows:—

1. Talc.	6. Orthoclase.
2. Gypsum.	7. Quartz.
3. Calcite.	8. Topaz.
4. Fluorite.	9. Corundum.
5. Apatite.	10. Diamond.

Arranged in this order, each member of the scale is harder than (*i. e.*, will scratch) all that come before it, and softer than (*i. e.*, is scratched by) all that come after it. The degree of hardness possessed by any mineral may be determined by direct comparison with the scale, and is expressed by the number, rather than the name, of the member of the scale to which it corresponds. Thus, if it does not scratch orthoclase, and is not distinctly scratched by orthoclase, its hardness is 6. If it scratches fluorite, but is scratched by apatite, its hardness is between 4 and 5 ; and it is possible, by making the test carefully, to determine whether it is about 4.25, 4.5 or 4.75.

Although it is important that the student should be acquainted with the scale of hardness, and understand how to use it, it is not essential for the determination of common minerals. Very few common minerals have hardness above 7 ; and hence for all ordinary purposes the thumb-nail, a knife or file and a piece of quartz are sufficient. In the column marked "H," in the tables, the exact, or where variable the average, hardness of each species is given, so that careful comparisons may be made when desired. But an examination of the analytical key on the left margin of each table, or of the general classification of the tables, will show that the determinations are based upon a scale of hardness embracing only five degrees, as follows :—

1. Very soft (below 2.5) ; can be scratched with the nail, or very easily with the knife.
2. Soft (2.5–4) ; cannot be scratched with the nail, but easily scratched with the knife.
3. Hard (4–6) ; can be be scratched with the knife, but not easily.
4. Very hard (6–7) ; cannot be scratched distinctly with the knife, but is scratched by quartz.
5. Adamantine (above 7) ; cannot be scratched by quartz.

The figures in the parentheses give the corresponding degrees of the regular scale.

The great advantage of this scale is its convenience, combined with a reasonable degree of accuracy. The adamantine minerals, or those having hardness above 7, are all found in one small group at the very end of the tables. Hence, in nearly all cases the quartz will not be required, the knife and thumb-nail being sufficient ; and a little practice will enable the student to determine whether a mineral which can be scratched with the knife scratches easily or with difficulty. The doubtful or variable minerals are repeated in all the sections of the tables to which students would be likely to assign them. It is important, however, to note that beginners usually overestimate the hardness of minerals; and that a sharp angle will usually slightly scratch a flat surface of the same mineral. The student will also learn to make allowance for foreign substances or impurities.

Tenacity.—All solid minerals may be classed as either *brittle* or *flexible*. Brittle minerals, to which class the majority belong, are those whose forms can not be sensibly distorted without rupture. The degree of cohesion, however, varies greatly; some brittle minerals being very rigid or strong and breaking with great difficulty, like corundum; while in the typical brittle minerals, such as quartz and calcite, the cohesion is less and a moderate blow suffices to break them. But the minimum cohesion is found in the friable minerals, which are at once very brittle and very easily broken. Brittle minerals, in other words, are those which break suddenly under a blow, and whether the blow needs be light or heavy is immaterial.

Flexible minerals, on the other hand, are those whose forms can be sensibly distorted without rupture. All flexible minerals are also *sectile*, *i. e.*, can be cut without breaking or crumbling. All brittle minerals are elastic, and the elastic flexible minerals, such as mica, are those possessing the least flexibility in pieces of ordinary thickness, constituting an intermediate class between the brittle and flexible minerals. The typical flexible minerals are malleable and ductile as well as sectile. As in the brittle class, the degree of cohesion varies greatly. This is seen by comparing iron and copper with gypsum and talc.

Properties Relating to Mass and Volume.—All that we have to consider here is the ratio of the mass and volume in minerals, *i. e.*, the

Specific Gravity.—By the specific gravity of a mineral we mean its weight compared with the weight of an equal volume of water. Although this property is quite as fundamental and constant as hardness, it is less useful in the determination of minerals, on account of the greater difficulty of measuring it with the same degree of accuracy. The specific gravity of solid bodies is usually determined as follows:—The specimen is first weighed carefully on a good balance; it is then suspended from one pan of the balance by a thread or fine wire in a glass of water, and while hanging freely and completely immersed, its weight is taken again. The second weight is subtracted from the first. The difference, or loss by immersion, is the weight of a volume of water equal to that of the specimen; and by dividing this weight into the first weight of the specimen, the desired ratio is obtained. For example, a piece of quartz weighs 25 grains in air, and 15.57 grains in water. $25 - 15.57 = 9.43$ grains, the weight of an equal volume of water. $25 \div 9.43 = 2.65$, the specific gravity of quartz.

Minerals exhibit a wide range in specific gravity, from petroleum, which floats on water, to gold, which is nearly twenty times heavier than water. Very few minerals, however, are heavier than iron (7.5) and still fewer are lighter than sulphur (2), the great majority falling between 2.5 and 5. But, notwithstanding this narrow range, the student can, with practice, learn, by merely lifting specimens to estimate the specific gravity accurately enough to make it a very valuable aid in the determination of many minerals. This property is relied upon to some extent, in the analytical key of the tables, for distinguishing groups of minerals. The caution with regard to impurities, given in the preceding section, might be repeated with additional emphasis here.

Properties Relating to Light.—On account of their comparatively superficial nature and the consequent ease with which they may be ascertained, the optical properties play a very important part in determinative mineralogy.

Lustre.—By the *lustre* or *glance* of a mineral is meant the quantity and quality of light reflected by it, as determined by the character or minute structure of its surface. Variations in the nature of the reflecting surface produce different *kinds* of lustre; and variations in the quantity of light reflected, *i. e.*, in the polish of the surface, produce different *degrees* of lustre.

The two principal kinds of lustre are the *metallic* and *non-metallic*. The metallic lustre is the lustre of all true metals, such as silver, copper, etc., and of most minerals in which metallic elements predominate. When the lustre is not distinctly or perfectly metallic, it is called *sub-metallic*. The *adamantine* lustre is intermediate between the metallic and non-metallic lustres. It is well shown in but few minerals, the diamond being the most perfect example.

Most of the minerals in which the non-metallic elements predominate have a *non-metallic* lustre. This is by far the most common lustre, only about one-fifth of the known minerals which have a distinct lustre being either metallic or adamantine. The non-metallic lustre embraces several varieties, which are named in the order of their importance :—

The *vitreous* is the lustre of glass, and of all minerals similar to glass in appearance, such as quartz, calcite, etc. The *resinous* lustre is seen in resins, of which the native mineral copalite is an example; it is also well exhibited in sulphur and sphalerite. The *pearly* lustre, *i. e.*, the lustre of pearl, is well shown only in minerals having a foliated or scaly structure, in other words, very perfect cleavage in one direction, such as talc, mica and gypsum. The *silky* or *satiny* lustre, like the pearly, is determined by the structure, being observed only in finely fibrous minerals. Fibrous gypsum or satin-spar is the best illustration, although fibrous calcite, serpentine, malachite, asbestus, etc., are nearly as good. The *greasy* and *waxy* lustres are most common in certain amorphous minerals, such as serpentine.

The degrees of lustre are expressed as follows :—

The lustre is *splendent* when the surface reflects brilliantly, giving well defined images. It is *shining* when the reflected image is not well defined. When there is a general reflection from the surface, but no recognizable image, the lustre is *glistening*. If the reflection is very imperfect, and chiefly from minute points, the lustre is *glimmering*. A mineral is described as *dull*, when there is a complete absence of lustre, *i. e.*, when no light is regularly reflected, as in earthy minerals.

In using these tables, or any determinative system, it is a matter of the first importance to be able to recognize the different kinds of lustre, and especially to distinguish the two principal kinds, metallic and non-metallic. This is the first question to be answered with every species. One aid to answering it correctly is found in the perfect opacity of all metallic minerals. If the finest splinter or the thinnest scale of the mineral in hand appears in the slightest degree translucent, when held up to the light, it cannot be metallic. But the converse statement is not always true, since some non-metallic minerals will appear quite opaque, except in the excessively thin sections prepared with considerable labor for microscopic examination. Metallic minerals, again, never grind to an earthy (chalk-like) powder, but when carefully examined the powder sparkles, the mineral showing its metallic shine or glance in the smallest particles.

The darker colored and nearly opaque non-metallic minerals are among those most likely to be referred to the wrong lustre ; and many of these, as well as the most of the sub-metallic, adamantine and dull species have, in consequence, been placed in both divisions of the tables.

Diaphaneity.—The light transmitted by minerals varies in amount within wide limits ; or, in other words, of the light received, and not reflected, more or less may be *absorbed*. The following degrees of of diaphaneity are usually recognized :—

Transparent, when the outline of an object seen through the mineral is perfectly distinct. *Subtransparent* or *semitransparent*, when objects are seen but the outlines are not distinct. *Translucent*, when light is transmitted but objects are not seen. *Subtranslucent* or *semitranslucent*, when merely the edges of the specimen transmit light or are translucent. *Opaque*, when the mineral transmits no light in ordinary specimens.

The diaphaneity is of very little value in the determination of minerals, except as an aid in distinguishing the metallic and non-metallic lustres ; for it is exceedingly variable and inconstant, the same species, or even the same specimen, often showing every degree from transparent to opaque. No other property of non-metallic minerals is affected in an equal degree by slight impurities and imperfections. In fact, if all mineral bodies were perfectly pure and homogeneous we should probably find that the metallic species are all opaque, and the non-metallic species are all transparent.

Color.—The color of a mineral is a measure of its power of absorbing different portions of the light that passes through it or falls on its surface. If no light is absorbed, or, more correctly, if all the rays of the spectrum are absorbed equally, the color is white. If the red rays are absorbed, the color is green, and so on. The color is always a mixture of the rays that are not absorbed.

The colors of metallic minerals are usually constant and afford important aid in distinguishing species; and the same is true of a few of the non-metallic minerals. The great majority of the non-metallic species, however, would be white if perfectly pure; and since the actual colors depend upon the kind and quantity of impurity present, they are extremely variable.

The different varieties of color in both metallic and non-metallic minerals are easily recognized; but a far more important distinction is that between *essential* and *non-essential* colors. By the essential color in any case is meant the color of the mineral itself in its purest state and in the finest particles. The non-essential colors, on the other hand, are chiefly the colors of the impurities contained in the minerals; although appearing in some cases to be due to the molecular structure. The essential color is known as the *streak*. Metallic minerals, which are always opaque, usually have essential colors, the color of the solid mineral and the powder being exactly or nearly the same. But non-metallic minerals, which are always somewhat diaphanous, usually have non-essential colors when not white, the color of the powder being white or gray regardless of the color of the solid mineral. The explanation is this: In opaque minerals we can only see the impurity immediately on the surface, and this is, as a rule, not sufficient to affect the color. But in diaphanous minerals we look *into* the specimen, and see impurity below the surface; and thus bring into view, in many cases, sufficient impurity so that its color drowns that of the pure mineral. To prove this we have only to take any mineral (serpentine is a good example) having a non-essential color and make it opaque by pulverizing it or abrading its surface. The non-essential color, the color of the impurity, immediately disappears just as water yellow with suspended clay becomes white when beaten into foam and thus made opaque. The essential color of a mineral, or the color of its powder, is called the *streak* of the mineral because the powder is most easily obtained by scratching the surface of the mineral and thereby pulverizing a minute portion of it. A still better method of observing the streak of all but adamantine minerals is to rub the specimen on an unpolished surface of white porcelain or Arkansas stone. The streak of very soft minerals is easily obtained by marking on paper, the lead-pencil mark being the streak of graphite. Very hard minerals must be pulverized in a mortar.

Any merely superficial color, or color due to exposure to the air or weather, is called *tarnish*. A mineral is described as tarnished when the surface color is different from that on a fresh fracture. The tarnish is often irised or marked by various prismatic colors, which are explained as due to interference of light caused by a film covering the mineral. This superficial film or layer may be the result of oxidation or other alteration of the mineral itself, or it may be a thin incrustation of some foreign substance.

Properties relating to Heat and Magnetism.—Under the head of heat, mineralogists investigate chiefly the *expansion* and *contraction* of minerals, their power of *absorbing, conducting* and *radiating* heat, and their *fusibility*.

Fusibility.—This is the most important of these properties and the only one that need be considered here. It is hardly practicable to determine with a thermometer the melting temperature of minerals, and the exact temperature of fusion is known for only a few species; but mineralogists are usually satisfied with knowing the relative fusing points of the different species. The relative fusibility is determined by comparison with six species which have been chosen as a *scale of fusibility*. These, beginning with the most readily fusible, are:—

1. Stibnite; 2. Natrolite; 3. Garnet; 4. Actinolite; 5. Orthoclase; 6. Bronzite; and we may add 7. Quartz. The fusibility of a mineral is determined by heating a fine splinter or fragment in the forceps or on charcoal with the oxidizing blowpipe flame. If it fuses as readily as natrolite, it is 2, if, like quartz, it cannot be even rounded on the thinnest edges, it is 7, and so on.

It is very important that the fusibility should be tested with very fine splinters or scales. When this precaution is observed, stibnite and natrolite are readily melted to liquid globules, and garnet with a stronger heat; but with actinolite, and still more with orthoclase and bronzite, the evidence of fusion is the rounding of the sharp edges of the splinters. A mere swelling up or intumescence of a mineral should not be mistaken for fusion.

Magnetism.—But few minerals are sensibly affected by an ordinary magnet; and these are the species containing the largest proportions of the strongly magnetic element—iron. The degree of magnetism is usually proportional to the percentage of iron, being greater in magnetite than in any other mineral except pure iron. Where the magnetism is weak, it can only be detected by first finely pulverizing the mineral.

Properties Relating to the Action of Minerals on the Senses.—These properties are of little importance, except with a few species.

Touch or Feel.—This is described as *meagre* (chalk, clay, etc.), *harsh, rough, smooth, unctuous* and *greasy*.

Taste.—A few minerals, being readily soluble, have a distinct taste. The principal kinds of taste mentioned are the *astringent, cooling, sour, bitter, saline* and *alkaline.*

Odor or Smell.—Solid minerals are usually devoid of odor unless subjected to some special treatment, as heating, rubbing, moistening, etc. The principal kinds of odor are the *sulphurous, arsenical, argillaceous* and *fetid.*

CHEMICAL PROPERTIES.

What we have to consider here, chiefly, is the characteristic chemical behavior of the common minerals, and especially of their principal constituent elements. A systematic account of the composition and chemical relations of minerals would be out of place in a work of this kind; but there is one general principle of such fundamental importance in determinative mineralogy as to demand the attention of the student at the very outset. This is the *relation of the composition to the physical properties of minerals.* The comparison of almost any mineral with the chemical elements of which it is composed shows that the properties of minerals are often very distinct from those of their constituent elements. Thus, the common mineral pyrite is composed of sulphur and iron, and in its general aspect it resembles neither of these elements; but a closer comparison shows that most of its properties may be observed in them; for it is yellow and brittle like sulphur, and metallic, hard and heavy like iron. Its specific gravity is almost an average of that of the two elements. In general, the contrast between minerals and their component elements is more strongly marked with the comparatively superficial properties, like color, than with the more fundamental properties, such as form, density, lustre, etc. The latter are, in a very large degree, the average of the properties of the elements. Thus the minerals in which heavy, metallic elements predominate, such as galenite, etc., are themselves heavy and have usually a metallic lustre; while minerals composed chiefly of light, non-metallic elements, such as quartz, etc., are characterized by a low specific gravity and non-metallic lustre. In accordance with this principle, we find that only those minerals are magnetic which are richest in iron, the magnetism increasing with the percentage of iron; and that, water being one of the lightest and softest of mineral-constituents, the hydrous species are lighter and softer than anhydrous species of otherwise similar composition.

General Blowpipe and Chemical Tests :—Certain standard tests recur so frequently in the tables that the student is recommended to acquire some familiarity with the *modus operandi* of each, and the general reactions which it yields, before proceeding to the more special tests for the indentification of the particular constituents of minerals.

Heating in the Closed Tube.—The mineral should be in the form of powder or fine grains; and the bulk of a kernel of wheat will be sufficient in most cases. This is placed in the bottom of a hard glass tube three inches long and closed at one end. This end of the tube may be heated by directing the blowpipe flame against it, or more easily by holding it in the flame of the alcohol lamp or Bunsen gas burner. The heat is applied gently at first, and then more strongly if the reaction seems to demand it. The most essential feature of this test is that the mineral is heated nearly out of contact with the air; and the principal changes or reactions observable are (1) fusion, and (2) the formation of a sublimate in the upper part of the tube. Few minerals are melted; and the only minerals yielding sublimates are those that are volatile or contain volatile constituents, such as sulphur, arsenic, mercury, water, etc.

Heating in the Open Tube.—The mineral, preferably in the form of one or several small grains, or of powder if it decrepitates, is placed in a tube of hard glass three inches long and open at both ends, the best position being three-fourths of an inch from one end; and this point is then heated in the same manner as the bottom of the closed tube, the tube being inclined as steeply as may be without having the mineral slide out. A current of air is thus caused to pass over the mineral while it is ignited, and the conditions are favorable for oxidation. The most important reactions are (1) the characteristic odors afforded by sulphides and arsenides, and (2) the sublimates formed in the upper part of the tube and consisting chiefly of the oxides of the same substances forming sublimates in the closed tube. Sulphurous and other acid vapors are given off which do not condense on the walls of the tube but are recognized by the reddening of a slip of moistened blue litmus paper inserted at the upper end of the tube.

Heating on Charcoal.—The mineral, in about the same quantity as before, is placed in a shallow, saucer-shaped cavity excavated near one end of a suitable piece of charcoal. The charcoal is held so that the flame can be directed upon the mineral with the blowpipe. If the mineral decrepitates or flies away, it may be finely pulverized and made into a paste with water, and then heated gradually until the mass coheres.

The principal phenomena to be observed are: (1) The odors of burning sulphur and arsenic, the latter resembling the odor of garlic. These are afforded by the elements and by sulphides and arsenides. (2) Fusion. (3) In the oxidizing flame, the metals are oxidized and characteristic sublimates or coatings of the oxides are formed on the charcoal, which may be recognized by their color, extent and degree of volatility. (4) In the reducing flame, several metals are readily reduced to the metallic state, especially if carbonate of soda is added to the pulverized minerals, and many compounds of iron will become magnetic.

Heating in the Forceps.—If possible, a minute, slender fragment or splinter should be held by its larger end with a fine point projecting well beyond the end of the forceps, so that it may be introduced into the tip of the blue or oxidizing flame without heating the forceps. If the mineral decrepitates so that it can not be held in the forceps, it may be finely pulverized, moistened with clean water, and supported on a loop of platinum wire.

The phenomena to be observed are: (1) The degree of fusibility of minerals. If the fine point melts into a round globule, it is readily fusible; if it is only slightly rounded on the end, it is difficultly fusible, and so on. For greater precision, the behavior of the mineral should be compared with that of similar splinters of the minerals forming the scale of fusibility. (2) The colors imparted to the flame by minerals. These are often very characteristic, and form a beautiful and delicate test for the following elements: All compounds of *soda* yield a bright *reddish yellow* flame. *Potash,* in most of its compounds, tinges the flame *bluish violet. Lithia* gives a *purplish red, lime* a *yellowish red* and *strontia* a *pure red,* color to the flame. *Baryta, copper, phosphoric acid* and *boric acid* give different shades of *green.* But the *chloride of copper* colors the flame *azure-blue.* (3) The colors imparted to the minerals themselves, when they are first moistened with a solution of nitrate of cobalt and then strongly heated. *Alumina compounds* become *blue, zinc oxide green* and *magnesia compounds flesh-red.*

Fusion with Borax and Salt of Phosphorus.—The tests with these reagents are most conveniently made on a short piece of platinum wire, one end of which is bent into a loop one-tenth of an inch in diameter. The loop is heated to redness and dipped into the pulverized flux; the portion which adheres to the wire is fused in the blowpipe flame; and this process is repeated until a full round bead of the flux is formed. When the bead is complete, and while it is still hot, it is touched to the pulverized mineral, so that a small portion of the latter adheres. The bead is then strongly heated, usually with the oxidizing flame, until the mineral, if soluble, is completely dissolved in the bead. These tests are especially adapted to the detection of the various metallic oxides; and the chief phenomenon to be observed is the color imparted to the bead or glass by the oxide.

Alkaline Reaction.—A fragment of the mineral is strongly heated in the forceps or on charcoal, and then placed upon a strip of red litmus paper and moistened with a drop of water. If the color of the paper is changed to blue, the reaction is alkaline. This is a valuable test for the alkalies and alkaline earths.

Dissolving in Acid.—Chlorhydric (muriatic) acid answers in most cases, nitric and sulphuric acids being rarely required. The acid should be used full strength, except when dilute acid is especially called for ; and it is very important, especially with silicates, that the mineral to be tested should be in the form of an *impalpable powder*. The behavior of minerals with acids is most conveniently observed in the test tube. A small quantity of the *pulverized* mineral is placed in the bottom of the tube, covered with one or two inches of acid, and heat applied gently until the acid boils.

List of the Principal Constituents of Minerals, with the Tests employed in the Tables for the Detection of Each.—The substances are arranged alphabetically, for convenience of reference ; and the student is expected to turn to this and the preceding section for full explanations of the tests indicated in the tables.

Alumina is recognized, in fusible minerals, by the beautiful blue color which they assume when strongly heated and then moistened with cobalt solution and heated again. Hard minerals must be pulverized

Antimony and most of its compounds fuse and yield dense white fumes of antimony oxide when heated on charcoal. The oxide partially condenses on the coal, forming an extended white coating, which is volatile in the blowpipe flame.

Arsenic and nearly all of its compounds yield a sublimate of metallic arsenic in the closed tube, and a white crystalline sublimate of arsenious acid in the open tube. On charcoal, this element is recognized by the abundant gray smoke, the strong garlic odor, and the extremely volatile gray coating which it forms on the coal.

Baryta is recognized by the yellowish green color imparted to the blowpipe flame by all of its compounds except silicates.

Bismuth fuses, volatilizes and oxidizes on charcoal, forming a yellow, volatile coating.

Boracic Acid colors the blowpipe flame an intense yellowish green. This test is more reliable if the mineral is moistened with sulphuric acid before heating.

Carbonic Acid is readily set free in the gaseous form, from all carbonates, by chlorhydric acid, especially if the minerel be pulverized or the acid heated. The effervescence of the escaping carbonic acid is distinguished from that due to sulphuretted hydrogen by the offensive odor of the latter gas.

Chlorine is most conveniently detected by adding the mineral to a salt of phosphorus bead which has been saturated with copper oxide, and observing the intense blue color which is imparted to the blowpipe flame.

Chromium is recognized by the emerald-green color which its oxide imparts to the borax bead.

Copper and its compounds impart a green color to the flame and to the borax and salt of phosphorus beads. In most cases the flame coloration will be blue if the mineral be first moistened with chlorhydric acid. Copper can be reduced from nearly all of its ores by fusion with soda on charcoal.

Fluorine is recognized in most of its compounds by heating the mineral in the closed tube with potassium bisulphate and pulverized glass. Fluorhydric acid is set free, which etches the glass in the bottom of the tube, and a white sublimate of silicon fluoride is formed in the upper part of the tube.

Iron may be recognized by the yellow and brown colors which its oxides impart to the borax bead; and by the fact that nearly all the compounds of iron become magnetic when heated, especially if heated on charcoal with soda.

Lead is reduced to the metallic state with soda, and forms a volatile yellow coating of lead oxide on the charcoal.

Lime colors the blowpipe flame yellowish red; and many compounds of lime give an alkaline reaction after heating.

Lithia usually colors the flame bright purplish red.

Magnesia is recognized in many of its compounds by the pink color which they assume when moistened with cobalt solution and strongly heated.

Manganese is recognized in all of its compounds by the highly characteristic colors which it imparts to the fluxes. With borax it forms a clear bead of a deep amethystine red or violet-red color; while fusion with soda (or, if necessary, with soda and nitre) yields an opaque, bright green mass.

Mercury always yields a sublimate of metallic mercury when any of its compounds are heated in a closed tube with soda. This sublimate consists of minute, liquid drops.

Phosphoric Acid may usually be recognized by the green color which phosphates impart to the flame, especially if previously moistened with sulphuric acid. Or the pulverized mineral may be fused in the closed tube with a bit of magnesium wire, and a drop of water added to the fused mass when cold, evolving phosphuretted hydrogen, which is recognized by its disagreeable odor.

Potassa may be often recognized by the violet color which it imparts to the flame.

Silica effervesces with soda on platinum wire, forming a clear glass, if the soda be not in excess. Most of the silicates are decomposed by salt of phosphorus, leaving a cloudy skeleton of silica floating in the bead. Many hydrous and basic silicates are decomposed by chlorhydric acid, leaving the silica in the form of a stiff jelly, or a floculent powder.

Silver is easily reduced with soda on charcoal, and forms a brown coating of silver oxide.

Soda is easily recognized by the intense yellow color which it imparts to the flame.

Strontia is known by its crimson flame. The color is more intense when the mineral is moistened with chlorhydric acid.

Sulphur, as it occurs in sulphides, may be recognized in many ways. Bisulphides, or those containing the largest proportion of sulphur, yield sublimates of sulphur in the closed tube. All sulphides redden moistened blue litmus paper in the open tube. If any sulphide is fused with soda in the closed tube and the fused mass transferred to a bright silver coin and moistened with water, a dark brown stain of silver sulphide is left on the coin. This test will also prove the presence of sulphur in any *sulphate*, if a little pulverized charcoal is added to the mixture in the closed tube before fusion, or the fusion is made on charcoal. All sulphides effervesce with hot chlorhydric acid, evolving sulphuretted hydrogen, which is recognized by its odor.

Tin is reduced to the metallic state with soda on charcoal.

Water may be expelled from nearly all hydrous minerals by heating in the closed tube, the vapor or steam condensing in the upper part of the tube and wetting the glass.

Zinc is reduced with soda on charcoal, but the metal immediately oxidizes, forming a coating on the coal which is yellow when hot and white when cold. The white and gray compounds of zinc assume a green color when moistened with cobalt solution and strongly ignited.

List of Apparatus and Reagents.

Blowpipe. The simple and inexpensive brass blowpipe used by jewellers answers very well for all the tests described in these tables; although the chemical blowpipe, with a platinum jet, a mouth-piece and a chamber for condensing the moisture of the breath, is, of course, rather more satisfactory in operation and more durable.

Fuel and Lamp. The most satisfactory source of heat for blowpipe and chemical experiments is the Bunsen gas-burner; the aperture for the admission of air at the bottom of the burner being closed by a sliding tube when the blowpipe is used, and left open when it is desired to use the direct heat of the flame. When gas is not available, a small alcohol lamp is recommended as most generally useful; although for use with the blowpipe alone a common oil lamp (olive oil is best) or stearine candle will answer in most cases. The blowpipe enables us to control not only the direction but the quality of the flame. The *reducing flame* (R. F.) is produced by holding the jet of the blowpipe outside of the flame and deflecting the entire flame by a gentle blast without essentially changing its character. The reducing flame should be used whenever it is desired to reduce to the metallic state, or partially deoxidize, metallic oxides. The *oxidizing flame* (O. F.) is obtained by introducing the jet into the flame and blowing more strongly, the deflected flame forming a slender blue cone. This is the flame usually employed, especially for oxidizing or roasting minerals, forming coatings on charcoal, and whenever a high temperature is desired, as in testing the fusibility and flame-coloration of minerals.

Forceps. Common steel forceps about four inches long will answer in most cases; although platinum-pointed forceps are very much better, if not essential, in testing fusibility and flame-coloration.

Platinum Wire. A piece about three inches long of moderately stout wire is the best. Before using, a circular loop one-tenth of an inch in diameter should be formed on the end of it.

Glass Tubes. These should be of hard glass, one-eighth to one-fourth inch in diameter, and three to four inches long. The *open* tubes are open at both ends, and are formed by simply cutting the tubing with a file into pieces of the proper length. The closed tubes are closed at one end, and are formed by heating pieces seven or eight inches long in the middle until the glass is soft enough to be drawn apart.

Charcoal. This important support should be made from soft wood and thoroughly burned, so as not to crack or snap when heated. Rectangular pieces four to eight inches long and two inches wide are the most convenient.

Agate Mortar. This instrument is very useful in reducing minerals to a fine powder, especially for treatment with acid; but it is unfortunately somewhat expensive.

Hammer, Anvil and Ring. These are employed to break and pulverize fragments of minerals, and answer in most cases as a substitute for the agate mortar.

File. A small three-cornered file is useful for testing the hardness and streak of minerals, and for cutting glass tubing. A notch is made with the file in one side of the tube, which is then gently bent and pulled apart, the nails of the two thumbs being brought together opposite the notch.

Magnet. A small magnet aids in the recognition of magnetic minerals.

Lens. A pocket lens or magnifying glass is useful in many ways.

Test-tubes Small test-tubes of hard glass are needed for treating minerals with acids.

Soda. (Carbonate of Sodium). **Borax** (Biborate of Sodium). **Salt of Phosphorus** (Phosphate of Sodium and Ammonium). **Oxide of Copper. Cobalt Solution** (Nitrate of Cobalt). **Chlorhydric, Sulphuric and Nitric Acids.**

Litmus Paper. The blue paper may be reddened by moistening it and holding it over the open mouth of the acid bottle.

SYNOPSIS OF THE CLASSIFICATION OF MINERALS.

SUBKINGDOM OF ELEMENTS.

Class of Metals.
1. *Gold Group.*—Gold, silver.
2. *Iron Group.*—Copper, iron, mercury.

Class of Metalloids.
3. *Arsenic Group.*—Bismuth, antimony, arsenic.
4. *Sulphur Group.*—Sulphur.
5. *Carbon-Silicon Group.*—Diamond, graphite.

SUBKINGDOM OF BINARY COMPOUNDS.

Class of Sulphides, Sulpharsenides, Arsenides, etc.
GOLD, IRON AND TIN SUBCLASS.
Proto or Galena Group
6. *Galena Family.*—Galenite, bornite, argentite.
7. *Blende Family.*—Sphalerite.
8. *Pyrrhotite Family.*—Cinnabar, millerite, pyrrhotite,
9. *Chalcocite Family.*—Chalcocite.   [niccolite.
Deuto or Pyrite Group.
10. *Pyrite Family.*—Pyrite, chalcopyrite, smaltite, cobaltite.
11. *Marcasite Family.*—Marcasite, arsenopyrite.
ARSENIC AND SULPHUR SUBCLASS.
12. *Realgar Family.*—Realgar.
13. *Orpiment Family.*—Orpiment, stibnite.
14. *Molybdenite Family.*—Molybdenite.
SULPHARSENITES, SULPHANTIMONITES, ETC.
15. Stephanite, tetrahedrite, pyrargyrite, proustite.

Class of Chlorides, Bromides and Iodides.
GOLD, IRON AND TIN SUBCLASS.
Anhydrous Division.
16. *Halite Family.*—Halite, cerargyrite.
Oxychloride Division.
17. *Atacamite Family.*—Atacamite.

Class of Fluorides.
GOLD, IRON AND TIN SUBCLASS.
Anhydrous Division.
18. *Fluorite Family.*—Fluorite.
19. *Cryolite Family.*—Cryolite.

Class of Oxides.
GOLD, IRON AND TIN SUBCLASS.
Anhydrous Division.
Protoxide Group.
20. *Cuprite Family.*—Cuprite.
21. *Zincite Family.*—Water, zincite.
22. *Melaconite Family.*—Melaconite.
Protoxide and Sesquioxide Group.
23. *Spinel Family.*—Spinel, magnetite, franklinite, chromite.
24. *Chrysoberyl Family.*—Chrysoberyl.

Sesquioxide Group.
    25.   *Corundum Family.*—Hematite, menaccanite, corundum.
Deutoxide Group.   [perofskite.
    26.   *Rutile Family.*—Cassiterite, rutile, zircon.
    27.   *Brookite Family.*—Brookite, pyrolusite.
Hydrous Division.
    28.   *Turgite Family.*—Turgite.
    29.   *Diaspore Family.*—Diaspore, göthite, manganite.
    30.   *Limonite Family.*—Limonite.
    31.   *Brucite Family.* Brucite, gibbsite.
    32.   *Psilomelane Family.*—Psilomelane, wad.
CARBON-SILICON SUBCLASS.
    33.   *Anhydrous Division.*—Quartz.
    34.   *Hydrous Division.*—Opal.

SUBKINGDOM OF TERNARY COMPOUNDS.

Class of Tantalates and Columbates.
    35.   *Tantalite Family.*—Tantalite, columbite, samarskite.
Class of Phosphates, Arsenates, etc.
Anhydrous Division.
    36.   *Apatite Family.*—Apatite, pyromorphite. mimetite,
Hydrous Division.   [vanadinite.
    37.   *Vivianite Family.*—Vivianite, erythrite.
    38.   *Wavellite Family.*—Lazulite, wavellite, turquois, autunite.
Class of Borates.
    39.   *Sassolite Family.*—Sassolite.
    40.   *Borax Family.*—Borax, ulexite.
Class of Tungstates, Molybdates and Chromates.
    41.   *Wolframite Family.*—Wolframite, wulfenite, scheelite.
Class of Sulphates.
Anhydrous Division.
    42.   *Celestite Family.*—Barite, celestite, anhydrite, anglesite.
Hydrous Division.
    43.   *Gypsum Family.*—Gypsum.
Class of Carbonates.
Anhydrous Division.
    44.   *Calcite Family.*—Calcite, dolomite, magnesite, siderite,
                          rhodochrosite, smithsonite.
    45.   *Aragonite Family.*—Aragonite, witherite, strontianite,
Hydrous Division.   [cerussite.
    46.   *Malachite Family.*—Malachite, azurite.
Class of Silicates.
Anhydrous Division.
  Bisilicate Group.
    47.   *Pyroxene Family.*—Enstatite, hypersthene, wollastonite,
                         pyroxene.
    48.   *Spodumene Family.*—Spodumene, rhodonite.
    49.   *Amphibole Family.*—Amphibole.
    50.   *Beryl Family.*—Beryl.
  Unisilicate Group.
    51.   *Phenacite Family.*—Willemite.
    52.   *Chrysolite Family.*—Chrysolite.
    53.   *Garnet Family.*—Garnet.
    54.   *Vesuvianite Family.*—Vesuvianite.
    55.   *Epidote Family.*—Epidote, allanite, zoisite.

56. *Mica Family.*—Muscovite, biotite, phlogopite, lepidom-
elane, lepidolite.
57. *Leucite Family.*—Sodalite, leucite, lapis-lazuli.
58. *Scapolite Family.*—Wernerite.
59. *Nephelite Family.*—Nephelite, cancrinite.
60. *Feldspar Family.*—Orthoclase, albite, oligoclase, labrad-
• orite, anorthite.
Subsilicate Group.
61. *Chondrodite Family.*—Chondrodite.
62. *Tourmaline Family.*—Tourmaline.
63. *Andalusite Family.*—Andalusite, fibrolite, cyanite.
64. *Euclase Family.*—Datolite, topaz.
65. *Titanite Family.*—Titanite.
66. *Staurolite Family.*—Staurolite.
Hydrous Division.
General Subdivision.
67. *Pectolite Family.*—Pectolite, laumontite.
68. *Chrysocolla Family.*—Chrysocolla.
69. *Calamine Family.*—Calamine, prehnite.
70. *Apophyllite Family.*—Apophyllite.
Zeolite Subdivision.
71. *Mesotype Family.*—Thomsonite, natrolite.
72. *Analcite Family.*—Analcite.
73. *Chabazite Family.*—Chabazite, gmelinite.
74. *Stilbite Family.*—Stilbite, heulandite.
Margarophyllite Subdivision.
75. *Talc Family.*—Talc, pyrophyllite.
76. *Sepiolite Family.*—Sepiolite, glauconite.
77. *Serpentine Family.*—Serpentine, deweylite.
78. *Kaolinite Family.*—Kaolinite.
79. *Pinite Family.*—Pinite.
80. *Hydromica Family.*—Hydromica.
81. *Chlorite Family.*—Ripidolite, prochlorite.
82. *Chloritoid Family.*—Margarite, chloritoid.

Hydrocarbons.
83. *Simple Hydrocarbons.*—Petroleum, etc.
84. *Oxygenated Hydrocarbons.*—Amber, copal, coal, etc.

To aid students in referring new minerals to their proper places
in the classification, and thus determining their relations to other
minerals, the name of each species in the tables is accompanied by
the number of the family to which it belongs.

GENERAL CLASSIFICATION.

Analytical Key.		Species.	Composition.	Lustre.	Color.	Streak.
	Micaceous. Foliæ elastic.	Phlogopite (56).	(KMgAl)$_2$ SiO$_4$.	Pearly to submetallic.	Yellow to brown.	Gray.
	G. 1.25.	Brown Coal (84).	C, H, O, etc.	Resinous to dull.	Brown to black.	Brown.
		Realgar (12).	AsS.	Resinous.	Aurora-red.	Aurora-red.
	G. 3–6.	Proustite (15).	Ag$_3$AsS$_3$.	Adamantine to dull.	Cochineal-red.	Cochineal-red.
		Pyrargyrite(15).	Ag$_3$SbS$_3$.	Adamantine to dull.	Black to cochineal-red.	Cochineal-red.
	G. 9, when pure.	Cinnabar (8). Compare *Red Ochre*	HgS.	Adamantine to dull.	Cochineal-red.	Scarlet.
	Micaceous. Foliæ elastic.	Phlogopite (56).	(KMgAl)$_2$ SiO$_4$.	Pearly to submetallic.	Yellow to brown.	Gray.
	Streak red. Malleable.	Copper (2).	Cu.	Metallic.	Copper-red.	Copper-red.
	Streak bright red. G. 9, when pure.	Cinnabar (8).	HgS.	Adamantine to dull.	Cochineal-red.	Scarlet.
		Proustite (15).	Ag$_3$AsS$_3$.	Adamantine to dull.	Cochineal-red.	Cochineal-red.
	Streak bright red.	Pyrargyrite(15).	Ag$_3$SbS$_3$.	Adamantine to dull.	Black to cochineal-red.	Cochineal-red.
	G. 5–6.	Cuprite (20).	Cu$_2$O.	Adamantine to dull.	Red to brown.	Red.
	Streak bright orange.	Zincite (21).	ZnO.	Sub-adamantine.	Red to orange.	Orange.
	Streak dull red to brown. G. 3–4.	Turgite (28).	2Fe$_2$O$_3$+ H$_2$O.	Submetallic to silky.	Reddish black to red.	Red.
		Sphalerite (7). Compare *Hematite.*	ZnS.	Resinous.	Brown.	Brown to yellow.

(Left margin rotated text: ¶. Color Red or Brown. — 1. Very Soft. — 2. Soft.)

H.	Tenacity.	G.	Form.	Cleavage.	Other Properties.	Confirmatory Chemical Tests.
2.75	Elastic and sectile.	2.8	IV. Foliated.	Basal, perfect.	Transparent to opaque.	Infusible; decomposed by strong sulphuric acid.
1.	Brittle to sectile.	1.25	Compact or laminated.	None.	Opaque.	Readily ignited, burning with flame.
2.	Brittle.	3.5	V. Also massive.	Clinopinacoidal and basal.	Transparent to translucent.	Volatile and combustible, burning with a blue flame and arsenical odor.
2.	Brittle.	5.5	III. Also massive.	Conchoidal.	Translucent.	Gives reactions for arsenic and silver.
2.	Brittle.	5.8	III. Also massive.	Conchoidal.	Translucent.	Gives reactions for antimony and silver.
2.	Brittle to sectile.	9.	III. Usually massive.	Uneven.	Usually opaque.	Volatile; with soda in closed tube yields sublimate of mercury.
2.75	Elastic and sectile.	2.8	IV. Foliated.	Basal, perfect.	Transparent to opaque.	Infusible; decomposed by strong sulphuric acid.
2.75	Malleable.	8.8	I. Also massive and arborescent.	None.	Opaque.	Fuses readily; green solution with nitric acid, which becomes blue with ammonia.
2.5	Brittle to sectile.	9.	III. Usually massive.	Uneven.	Usually opaque.	Volatile; with soda in closed tube yields sublimate of mercury.
2.5	Brittle.	5.5	III. Also massive.	Conchoidal.	Translucent to opaque.	Gives reactions for arsenic and silver.
2.5	Brittle.	5.8	III. Also massive.	Conchoidal.	Translucent to opaque.	Gives reactions for antimony and silver.
3.5	Brittle.	6.	I. Also capillary and massive.	Octahedral.	Translucent to opaque.	Colors flame green and fuses readily, yielding metallic copper.
4.	Brittle.	5.5	III. Also massive.	Basal, perfect.	Usually opaque.	Infusible; soluble in acids; zinc coating with soda on charcoal.
5.	Brittle.	3.7	Massive, botryoidal, fibrous, etc.	Uneven, splintery, etc.	Opaque.	Yields water in closed tube; magnetic on charcoal.
3.5	Brittle.	4.	I. Also massive.	Dodecahedral, perfect.	Transparent to opaque.	Reactions for sulphur; zinc coating with soda on charcoal; effervesces in acid (H_2S).

Analytical Key.			Species.	Composition.	Lustre.	Color.	Streak.
I. Color Red or Brown.—Continued.	*2. Soft.— Continued.*	Streak grayish black.	Pyrrhotite (8).	Fe_7S_8.	Metallic.	Bronze-yellow to copper-red.	Dark grayish black.
			Bornite (6).	(FeCu)S.	Metallic.	Dark copper-red.	Pale grayish black.
	3. Hard.	Streak black.	Niccolite (8).	NiAs.	Metallic.	Pale copper-red.	Brownish black.
			Pyrrhotite (8).	Fe_7S_8.	Metallic.	Bronze-yellow to copper-red.	Grayish black.
		Streak gray to brown.	Sphalerite (7).	ZnS.	Resinous.	Brown.	Brown.
			Brookite (27). Compare *Cuprite*.	TiO_2.	Adamantine.	Brown.	Pale brown.
		Streak yellow or orange.	Zincite (21).	ZnO.	Sub-adamantine.	Red to orange.	Orange.
			Göthite (29).	$Fe_2O_3 + H_2O$.	Adamantine to dull.	Brown.	Yellow.
			Limonite (30).	$2Fe_2O_3 + 3H_2O$.	Submetallic to silky.	Brown.	Yellow.
		Streak red.	Cuprite (20).	Cu_2O.	Adamantine to dull.	Red to brown.	Red.
			Turgite (28).	$2Fe_2O_3 + H_2O$.	Submetallic to silky.	Reddish black to red.	Red.
			Hematite (25).	Fe_2O_3.	Metallic to dull.	Black to red.	Red.
	4. Very Hard.	G. nearly 7.	Cassiterite (26).	SnO_2.	Adamantine.	Brown.	Gray to brown.
		G. 4.2	Rutile (26).	TiO_2.	Adamantine.	Brown.	Gray to yellowish.
			Brookite (27).	TiO_2.	Adamantine.	Brown.	Pale brown.

H.	Tenacity.	G.	Form.	Cleavage.	Other Properties.	Confirmatory Chemical Tests.
3.5	Brittle.	4.5	III. Usually massive.	Uneven.	Opaque. Tarnishes.	Fusible to a black, magnetic mass; usually slightly magnetic before fusion.
3.	Brittle.	5.	I. Usually massive.	Octahedral, in traces.	Opaque Blue and green tarnish.	Gives reactions for copper; fuses to a black, magnetic globule.
5.25	Brittle.	7.5	III. Usually massive.	Uneven.	Opaque.	Arsenic fumes and coating and magnetic globule on charcoal.
4.5	Brittle.	4.5	III. Usually massive.	Uneven.	Opaque.	Fusible to a black, magnetic mass; usually slightly magnetic before fusion.
4.	Brittle.	4.	I. Usually massive.	Dodecahedral, perfect.	Transparent to opaque.	Reactions for sulphur; zinc coating with soda on charcoal; effervesces in acid (H_2S).
5.5	Brittle.	4.2	IV. Small square crystals.	Prismatic, indistinct.	Translucent to opaque.	Infusible and insoluble.
4.5	Brittle.	5.5	III. Also massive.	Basal, perfect.	Usually opaque.	Infusible; soluble in acid; zinc coating with soda on charcoal.
5.25	Brittle.	4.2	IV. Usually massive, fibrous or botryoidal.	Prismatic, perfect.	Opaque.	Water in closed tube; magnetic on charcoal.
5.25	Brittle.	4.	Massive, botryoidal, fibrous, etc.	Uneven, splintery, etc.	Opaque.	Like göthite.
4.	Brittle.	6.	I. Also capillary and massive.	Octahedral.	Translucent to opaque.	Colors flame green and fuses readily, yielding metallic copper.
5.5	Brittle.	3.7	Massive, botryoidal, fibrous, etc.	Uneven, splintery, etc.	Opaque.	Water in closed tube; magnetic on charcoal.
5.5	Brittle.	4.5	Massive, botryoidal, fibrous, etc.	Uneven, splintery, etc.	Opaque.	No water in closed tube; magnetic on charcoal.
6.5	Brittle.	6.8	II. Also massive, botryoidal, etc.	Prismatic, indistinct.	Nearly opaque.	Infusible and insoluble; with soda on charcoal reduced to metallic tin.
6.25	Brittle.	4.2	II. Also massive.	Prismatic, distinct.	Nearly opaque.	Infusible and insoluble.
6.	Brittle.	4.2	IV. Small square crystals.	Prismatic, indistinct.	Translucent to opaque.	Like rutile.

Analytical Key.			Species.	Composition.	Lustre.	Color.	Streak.
II. Color Yellow.	*1. Very Soft.*	Malleable.	Gold (1).	Au.	Metallic.	Gold-yellow.	Yellow.
		Micaceous. Foliæ elastic.	Phlogopite (56).	(KMgAl)₂ SiO₄.	Pearly to submetallic.	Yellow to brown.	Gray.
		Brittle.	Orpiment (13).	As₂S₃.	Pearly to resinous.	Lemon-yellow.	Yellow.
	2. Soft.	Streak yellow. Malleable.	Gold (1).	Au.	Metallic.	Gold-yellow.	Yellow.
		Streak yellow. Brittle.	Sphalerite (7).	ZnS.	Resinous.	Brown to yellow.	Dull yellow.
			Millerite (8).	NiS.	Metallic.	Brass-yellow.	Bright yellow.
			Zincite (21).	ZnO.	Sub-adamantine.	Red to orange.	Orange.
		Streak black or nearly so.	Chalcopyrite (10).	(CuFe)S₂.	Metallic.	Brass-yellow.	Greenish black.
			Pyrrhotite (8).	Fe₇S₈.	Metallic.	Bronze-yellow.	Grayish black.
			Bornite (6).	(CuFe)S.	Metallic.	Brownish copper-red.	Grayish black.
	3. Hard.	Streak nearly black. H. about 4.	Chalcopyrite (10).	(CuFe)S₂.	Metallic.	Brass-yellow.	Greenish black.
			Pyrrhotite (8).	Fe₇S₈.	Metallic.	Bronze-yellow.	Grayish black.
		Streak nearly black. II. about 6.	Marcasite (11).	FeS₂.	Metallic.	Grayish yellow.	Grayish black.
			Pyrite (10).	FeS₂.	Metallic.	Pale brass-yellow.	Greenish black.
			Arsenopyrite (11).	FeAsS.	Metallic.	White to gray.	Dark grayish black.

H.	Tenacity.	G.	Form.	Cleavage.	Other Properties.	Confirmatory Chemical Tests.
2.5	Malleable.	19.3	I. Usually in grains or nuggets.	None.	Opaque.	Readily fusible; insoluble in common acids; does not tarnish.
2.75	Elastic and sectile.	2.8	IV. Foliated.	Basal, perfect.	Transparent to opaque.	Infusible; decomposed by strong sulphuric acid.
1.75	Sectile to brittle.	3.5	IV. Foliated and massive.	Macropinacoidal, very perfect.	Translucent.	Volatilizes and gives reactions for arsenic and sulphur.
3.	Malleable.	19.3	I. Usually in grains or nuggets	None.	Opaque.	Readily fusible; insoluble in common acids; does not tarnish.
3.5	Brittle.	4.	I. Usually massive.	Dodecahedral, perfect.	Translucent to opaque.	Reactions for sulphur; with soda on charcoal gives a zinc coating.
3.25	Brittle.	5.	III. Acicular or fibrous.	Rhombohedral.	Opaque.	Reactions for sulphur; fuses on charcoal to a magnetic globule.
4.	Brittle.	5.5	III. Also massive.	Basal, perfect.	Usually opaque.	Infusible; soluble in acids; zinc coating with soda on charcoal.
3.5	Brittle.	4.2	II. Usually massive.	Uneven.	Opaque. Iridescent tarnish.	Sulphur in closed tube; magnetic globule on charcoal; green solution in nitric acid.
3.5	Brittle.	4.5	III. Usually massive.	Uneven.	Opaque.	Fusible to magnetic mass; usually slightly magnetic before fusion.
3.	Brittle.	5.	I. Usually massive.	Uneven.	Opaque. Blue and green tarnish.	Reactions for copper; fuses to magnetic globule.
4.	Brittle.	4.2	II. Usually massive.	Uneven.	Opaque. Iridescent tarnish.	Sulphur in closed tube; magnetic globule on charcoal; green solution in nitric acid.
4.5	Brittle.	4.5	III. Usually massive.	Uneven.	Opaque. Tarnishes.	Fusible to magnetic mass; usually slightly magnetic before fusion.
6.	Brittle.	4.7	IV. Often massive or globular.	Prismatic.	Opaque. Decomposes readily.	Sulphur in closed tube; magnetic residue on charcoal.
6.	Brittle.	5.	I. Cubes and pyritohedrons; often massive.	Uneven.	Opaque.	Like marcasite.
5.5	Brittle.	6.2	IV. Also massive.	Prismatic, distinct.	Opaque. Tarnishes.	Arsenic coating and magnetic residue on charcoal; arsenic sublimate in tube.

Analytical Key.			Species.	Composition.	Lustre.	Color.	Streak.
II. *Color Yellow.—Continued.*	*3. Hard.—Continued.*	Streak yellow. Crystalline or granular.	Sphalerite (7).	ZnS.	Resinous.	Brown to yellow.	Dull yellow.
			Zincite (21).	ZnO.	Sub-adamantine.	Red to orange.	Orange.
		Streak yellow. Compact or fibrous.	Limonite (30).	$2Fe_2O_3 + 3H_2O$.	Submetallic to silky.	Brown to yellow.	Yellow.
			Göthite (29).	$Fe_2O_3 + H_2O$.	Imperfect adamantine.	Brown to yellow.	Yellow.
	4. Very Hard.	Grayish yellow.	Marcasite (11).	FeS_2.	Metallic.	Grayish yellow.	Grayish black.
			Arsenopyrite (11).	FeAsS.	Metallic.	White to gray.	Dark grayish black.
		Pale brass-yellow.	Pyrite (10).	FeS_2.	Metallic.	Pale brass-yellow.	Greenish black.
III. *Color Black.*	*1. Very Soft.*	Micaceous. Foliæ elastic.	Biotite (56).	(AlMgK Fe)$_2$ SiO$_4$.	Pearly to submetallic.	Black.	Gray.
		Greasy feel. Usually foliated.	Graphite (5).	C.	Metallic.	Iron-black to dark gray.	Black.
			Molybdenite (14).	MoS_2.	Metallic.	Lead-gray.	Lead-gray.
		Malleable and sectile.	Argentite (6).	AgS.	Metallic.	Backish lead-gray.	Blackish lead-gray.
		Streak black.	Wad (32).	$2MnO_2 + H_2O$.	Dull.	Black.	Black.
			Pyrolusite (27).	MnO_2.	Metallic.	Iron-black to dark gray.	Black.
			Stephanite (15).	Ag_5SbS_3.	Metallic.	Iron-black.	Iron-black.
		Streak red.	Pyrargyrite (15).	Ag_3SbS_3.	Adamantine.	Black to cochineal-red.	Cochineal-red.

H.	Tenacity.	G.	Form.	Cleavage.	Other Properties.	Confirmatory Chemical Tests.
4.	Brittle.	4.	I. Usually massive.	Dodecahedral, perfect.	Translucent to opaque.	Reactions for sulphur; zinc coating with soda on charcoal; effervesces in acid (H_2S).
4.5	Brittle.	5.5	III. Also massive.	Basal, perfect.	Usually opaque.	Infusible; soluble in acids; zinc coating with soda on charcoal.
5.25	Brittle.	4.	Massive, botryoidal, fibrous, etc.	Uneven, splintery, etc.	Opaque.	Water in closed tube; magnetic residue on charcoal.
5.25	Brittle.	4.2	IV. Massive, fibrous or botryoidal.	Prismatic, perfect.	Opaque.	Like limonite.
6.5	Brittle.	4.7	IV. Often massive or globular.	Prismatic.	Opaque. Decomposes readily.	Sulphur in closed tube; magnetic residue on charcoal.
6.	Brittle.	6.2	IV. Also massive.	Prismatic, distinct.	Opaque. Tarnishes.	Arsenic coating and magnetic residue on charcoal; arsenic sublimate in tube.
6.5	Brittle.	5.	I. Cubes, etc.; also massive.	Uneven.	Opaque.	Sulphur in closed tube; magnetic residue on charcoal.
2.5	Elastic and sectile.	2 9	III. Foliated.	Basal, perfect.	Transparent. Green by transmitted light.	Infusible; decomposed by strong sulphuric acid.
1.25	Sectile.	2.2	III. Usually foliated.	Basal, perfect.	Opaque. Greasy feel.	Infusible; unaltered by acids.
1.25	Sectile.	4.6	III. Usually foliated.	Basal, perfect.	Opaque. Greasy feel.	Infusible; reactions for sulphur.
2.	Malleable and sectile.	7.3	I. Also massive.	Uneven.	Opaque.	Reactions for sulphur; silver on charcoal.
1.	Earthy.	3.8	Amorphous.	Earthy.	Opaque.	Water in closed tube; amethystine bead with borax; evolves chlorine with HCl.
2.	Brittle.	4.8	IV. Usually columnar or massive.	Prismatic.	Opaque.	Infusible; amethystine bead with borax; evolves chlorine with HCl.
2.	Brittle.	6.25	IV. Usually massive.	Imperfect.	Opaque.	On charcoal gives antimony coating and globule of silver.
2.	Brittle.	5.8	III. Also massive.	Conchoidal.	Translucent.	Gives reactions for antimony and silver.

Analytical Key.			Species.	Composition.	Lustre.	Color.	Streak.
III. Color Black.—Continued.	*2. Soft.*	Micaceous.	**Biotite** (56).	(KFeMg Al)$_2$SiO$_4$.	Pearly to submetallic.	Black.	Gray.
			Lepidomelane (56).	(KFeAl)$_2$ SiO$_4$.	Adamantine to pearly.	Black.	Grayish green.
		Malleable.	**Argentite** (6).	AgS.	Metallic.	Blackish lead-gray.	Blackish lead-gray.
		Streak red.	**Hematite** (25).	Fe$_2$O$_8$.	Metallic.	Black to gray.	Red.
			Pyrargyrite (15).	Ag$_3$SbS$_8$.	Adamantine to dull.	Black to cochineal-red.	Cochineal-red.
		Streak brown.	**Sphalerite** (7).	ZnS.	Submetallic.	Black.	Brown.
		Streak white or gray.	**Arsenic** (3).	As.	Metallic.	White.	White or gray.
		Streak black. G. 1 to 2.	**Mineral Coal** (84).	C,H,O, etc.	Submetallic.	Black.	Black.
		Streak black or nearly so. G. 4 to 6.	**Wad** (32).	2MnO$_2$ +H$_2$O.	Dull.	Black.	Brownish black.
			Manganite (29).	Mn$_2$O$_8$ +H$_2$O.	Submetallic.	Iron-black to steel-gray.	Brownish black.
			Pyrolusite (27).	MnO$_2$.	Metallic.	Iron-black to dark gray.	Black.
			Chalcocite (9).	CuS.	Metallic.	Blackish lead-gray.	Blackish lead-gray.
			Melaconite (22).	CuO.	Metallic to dull.	Black.	Black.
			Tetrahedrite (15).	Cu$_8$Sb$_2$S$_7$	Metallic.	Dark gray to iron-black.	Dark gray to black.
			Stephanite (15).	Ag$_6$SbS$_4$.	Metallic.	Black.	Black.

H.	Tenacity.	G.	Form.	Cleavage.	Other Properties.	Confirmatory Chemical Tests.
2.75	Elastic and sectile.	2.9	III. Foliated.	Basal, perfect.	Transparent. Green by transmitted light.	Infusible; decomposed by strong sulphuric acid.
3.	Sectile to brittle.	3.	III. Foliated.	Basal, perfect.	Opaque to translucent.	Fuses to magnetic globule; decomposed by sulphuric acid.
2.5	Malleable and sectile.	7.3	I. Also massive.	Uneven.	Opaque.	Reactions for sulphur; silver on charcoal.
4.	Brittle.	4.9	III. Scaly or foliated.	Basal.	Opaque. Sometimes magnetic.	Infusible; magnetic on charcoal; soluble in HCl.
2,5	Brittle.	5.8	III. Also massive.	Conchoidal.	Translucent to opaque.	Gives reactions for antimony and silver.
3.5	Brittle.	4.	I. Usually massive	Dodecahedral, perfect.	Opaque.	Reactions for sulphur; zinc coating with soda on charcoal; effervesces in acid (H_2S).
3.5	Brittle.	6.	III. Massive or botryoidal	Basal, imperfect.	Opaque. Dark gray tarnish.	Volatilizes without fusing; gray fumes and coating on charcoal; metallic sublimate in closed tube.
2.5	Brittle.	1.5	Compact to laminated.	Even to conchoidal.	Opaque.	Burns before blowpipe; infusible and insoluble in acids.
3.	Earthy to brittle.	3.8	Amorphous.	Uneven.	Opaque.	Water in closed tube; amethystine bead with borax; evolves chlorine with HCl.
4.	Brittle.	4.3	IV. Usually acicular or columnar.	Prismatic, perfect.	Opaque.	Water in closed tube; amethystine bead with borax.
2.5	Brittle.	4.8	IV. Usually columnar or massive.	Prismatic.	Opaque.	Infusible; amethystine bead with borax; evolves chlorine with HCl.
2.75	Brittle.	5.7	IV. Often massive.	Prismatic.	Opaque.	Fusible; blue flame with HCl; copper with soda on charcoal; sulphur reaction.
3.	Brittle to earthy.	6.25	IV. Usually massive or earthy.	Uneven or earthy.	Opaque.	Infusible; copper with soda on charcoal; green solution with nitric acid.
3.	Brittle.	4.8	I. Usually massive.	Uneven.	Opaque.	Reactions for antimony; copper with soda on charcoal.
2.5	Brittle.	6.25	IV. Usually massive.	Imperfect.	Opaque.	Reactions for antimony; silver with soda on charcoal.

Analytical Key.		Species.	Composition.	Lustre.	Color.	Streak.
		Strongly magnetic. Malleable.	Iron (2).	Fe.	Metallic.	Iron-gray to black.	Iron-gray to black.
		Strongly magnetic. Brittle.	**Magnetite** (23).	Fe_3O_4.	Metallic.	Iron-black.	Black.
		Streak black. G. above 7.	**Wolframite** (41).	$(FeMn)\ WO_4$.	Submetallic.	Brownish black.	Brownish black.
		Streak black. Uncrystalline.	**Psilomelane** (32).	$MnO_2 + H_2O$.	Submetallic.	Iron-black.	Brownish black.
			Menaccanite (25).	$(FeTi)_2O_3$.	Submetallic.	Iron-black.	Brownish black.
		Streak black. Crystalline.	**Manganite** (29).	$Mn_2O + H_2O$.	Submetallic.	Iron-black to steel-gray.	Brownish black.
			Columbite (35).	$FeCb_2O_6$.	Submetallic.	Iron-black.	Brownish black.
		Streak dark gray.	**Brookite** (27).	TiO_2.	Metallic to adamantine.	Black.	Dark gray.
			Allanite (55).	Complex silicate.	Submetallic.	Black.	Dark gray.
		Streak brown. H. 4.	**Sphalerite** (7).	ZnS.	Resinous to adamantine.	Black.	Brown.
			Chromite (23).	$FeCr_2O_4$.	Submetallic.	Iron-black.	Grayish brown.
		Streak brown. H. 5.5.	**Franklinite** (23).	$(FeZnMn)_3 O_4$.	Metallic.	Iron-black.	Reddish brown.
			Samarskite (35).	Complex columbate.	Submetallic, shining.	Black.	Reddish brown.
			Hematite (25).	Fe_2O_3.	Metallic to dull.	Iron-black. to steel-gray.	Red.
		Streak red.	**Turgite** (28).	$2Fe_2O_3 + H_2O$.	Submetallic.	Reddish black.	Red.

III. Color Black.—Continued.   *3. Hard.*

H.	Tenacity.	G.	Form.	Cleavage.	Other Properties.	Confirmatory Chemical Tests.
4.5	Malleable.	7.5	I. Usually compact.	Octahedral; usually none.	Opaque. Strongly magnetic.	Infusible; soluble in HCl,
5.5	Brittle.	5.	I. Usually granular.	Octahedral.	Opaque. Strongly magnetic.	Infusible; soluble in HCl.
5.5	Brittle.	7.4	V. Also lamellar or massive.	Prismatic, perfect.	Opaque.	Fuses easily to a magnetic globule; reactions for manganese.
5.	Brittle.	4.2	Compact and often botryoidal.	Even or conchoidal.	Opaque.	Infusible; water in closed tube; evolves chlorine with HCl.
5.	Brittle.	4.75	III. Often laminated.	None.	Opaque. Often slightly magnetic.	Infusible; reactions for iron.
4.	Brittle.	4.3	IV. Usually prismatic or columnar.	Prismatic, perfect.	Opaque. Never magnetic.	Infusible; water in closed tube; amethystine bead with borax.
6.	Brittle.	6.	IV. Crystals to coarsely massive.	Prismatic.	Opaque.	Infusible.
5.5	Brittle.	4.2	IV. Square crystals.	Prismatic, indistinct.	Opaque.	Infusible and insoluble.
5.5	Brittle.	3.6	V. Tabular and prismatic.	In traces.	Opaque.	Fuses with intumescence to a magnetic globule; gelatinizes with HCl.
4.	Brittle.	4.	I. Usually massive.	Dodecahedral, perfect.	Opaque.	Reactions for sulphur; zinc coating with soda on charcoal.
5.5	Brittle.	4.4	I. Usually massive.	None.	Opaque. Sometimes magnetic.	Infusible; green bead with borax.
5.5	Brittle	5.	I. Octahedrons, also massive.	Indistinct.	Opaque. Slightly magnetic.	Infusible; reactions for manganese and zinc with borax and soda.
5.5	Brittle.	5.7	V. Also massive.	None.	Opaque.	Emerald-green bead with SPh; green bead with soda.
5.5	Brittle.	4.9	III. Scaly, also massive or botryoidal.	Basal.	Opaque. Sometimes magnetic.	Infusible; magnetic on charcoal; soluble in HCl.
5.	Brittle.	3.7	Massive and botryoidal.	None.	Opaque.	Like hematite, but yields water in closed tube.

Analytical Key.			Species.	Composition.	Lustre.	Color.	Streak.
III. Color Black.—Continued.	*3. Hard.—Continued.*	Streak yellow.	Göthite (29).	Fe_2O_3 $+H_2O$.	Adamantine, imperfect.	Brownish black.	Yellow.
			Limonite (30).	$2Fe_2O_3$ $+3H_2O$.	Submetallic to silky.	Brown to black.	Yellow.
	4. Very Hard.	Strongly magnetic.	Magnetite (23).	Fe_3O_4.	Metallic.	Iron-black.	Black.
			Emery (25).	Al_2O_3 $+Fe_3O_4$.	Metallic.	Black to brown.	Black to brown.
		Streak black. Crystalline.	Menaccanite (25).	$(FeTi)_2O_3$.	Submetallic.	Iron-black.	Brownish black.
			Columbite (35).	$FeCbO_6$.	Submetallic.	Iron-black.	Brownish black.
		Streak black. Uncrys-talline.	Psilomelane (32).	MnO_2 $+H_2O$.	Submetallic.	Iron-black.	Brownish black.
		Streak gray or light brown. G. about 7.	Cassiterite (26).	SnO_2.	Adamantine.	Black.	Gray to light brown.
		Streak gray or light brown. G. about 4.	Rutile (26).	TiO_2.	Metallic to adamantine.	Black.	Gray to light brown.
			Brookite (27).	TiO_2.	Metallic to adamantine	Black.	Gray to light brown.
			Allanite (55).	Complex silicate.	Submetallic.	Black.	Gray to light brown.
		Streak dark reddish brown.	Franklinite (23).	$(FeZnMn)_3$ O_4.	Metallic.	Iron-black.	Reddish brown.
			Samarskite (35).	Complex columbate.	Submetallic, shining.	Black.	Reddish brown.
		Streak red.	Hematite (25).	Fe_2O_3.	Metallic.	Iron-black to steel-gray.	Red.
			Turgite (28). Compare *Limonite*.	$2Fe_2O_3$ $+H_2O$.	Submetallic.	Reddish black.	Red.

H.	Tenacity.	G.	Form.	Cleavage.	Other Properties.	Confirmatory Chemical Tests.
5.	Brittle.	4.2	IV. Columnar, fibrous and botryoidal.	Prismatic, perfect.	Opaque.	Water in tube; magnetic on charcoal.
5.	Brittle.	3.8	Massive, fibrous and botryoidal.	None.	Opaque.	Like göthite.
6.5	Brittle.	5.	I. Usually massive.	Octahedral.	Opaque. Strongly magnetic.	Infusible; soluble in HCl.
7to9	Brittle.	4.5	III. Coarsely to finely granular.	None.	Opaque. Strongly magnetic.	Infusible and insoluble.
6.	Brittle.	4.75	III. Often laminated.	None.	Opaque. Often slightly magnetic.	Infusible; reactions for iron.
6.	Brittle.	6.	IV. Usually in crystals.	Prismatic.	Opaque.	Infusible
6.	Brittle.	4.2	Compact or botryoidal.	Even or conchoidal.	Opaque.	Infusible; water in closed tube; evolves chlorine with HCl.
6.5	Brittle.	6.8	II. Also massive and botryoidal.	Imperfect.	Opaque.	Infusible; metallic tin with soda on charcoal.
6.25	Brittle.	4.2	II. Prismatic and twin crystals.	Prismatic, distinct.	Opaque.	Infusible.
6.	Brittle.	4.2	IV. Square crystals.	Prismatic, indistinct.	Opaque.	Like rutile.
6.	Brittle.	3.6	V. Tabular and prismatic.	In traces.	Opaque.	Fuses with intumescence to a magnetic globule; gelatinizes with HCl.
6.5	Brittle.	5.	I. Octahedrons and massive.	Indistinct.	Opaque. Slightly magnetic.	Infusible; reactions for manganese and zinc with borax and soda.
6.	Brittle.	5.7	IV. Also massive.	None.	Opaque.	Emerald-green bead with SPh; green bead with soda.
6.5	Brittle.	4.9	III. Scaly, also compact and botryoidal.	Basal.	Opaque. Sometimes magnetic.	Infusible; magnetic on charcoal; soluble in HCl.
6.	Brittle.	3.7	Compact, botryoidal and earthy.	None.	Opaque.	Like hematite; but yields water in closed tube.

Analytical Key.		Species.	Composition.	Lustre.	Color.	Streak.
I. Very Soft.	Greasy feel. Usually foliated.	Graphite (5).	C.	Metallic.	Black to dark gray.	Black.
		Molybdenite (14).	MoS_2.	Metallic.	Lead-gray.	Lead-gray.
	Malleable and sectile.	Argentite (6).	AgS.	Metallic.	Blackish lead-gray.	Blackish lead-gray.
	Streak black.	Pyrolusite (27).	MnO_2.	Metallic.	Iron-black to gray.	Black.
	Streak gray.	Stibnite (13).	Sb_2S_3.	Metallic.	Light lead-gray.	Lead-gray.
		Bismuth (3).	Bi.	Metallic.	Reddish white.	White to gray.
IV. Color Gray. *2. Soft.*	Malleable and sectile.	Argentite (6).	AgS.	Metallic.	Blackish lead-gray.	Blackish lead-gray.
	Streak black. Color bronze-yellow.	Pyrrhotite (8).	Fe_7S_8.	Metallic.	Bronze-yellow.	Grayish black.
	Streak black. Color steel-gray to iron-black.	Manganite (29).	$Mn_2O_3 +H_2O$.	Submetallic.	Iron-black to steel-gray.	Brownish black.
		Pyrolusite (27).	MnO_2.	Metallic.	Iron-black to steel-gray.	Black.
		Tetrahedrite (15).	$Cu_8Sb_2S_7$.	Metallic.	Dark gray to iron-black.	Dark gray to black.
		Melaconite (22).	CuO.	Metallic to dull.	Black to gray.	Black.
	Streak gray. Color white or light gray on fresh surface.	Arsenic (3).	As.	Metallic.	Tin-white.	Gray.
		Antimony (3).	Sb.	Metallic.	Tin-white.	Gray.
		Bismuth (3).	Bi.	Metallic.	Reddish white.	White to gray.

H.	Tenacity.	G.	Form.	Cleavage.	Other Properties.	Confirmatory Chemical Tests.
1.25	Sectile.	2.2	III. Usually foliated.	Basal, perfect.	Opaque. Greasy feel.	Infusible; unaltered by acids.
1.25	Sectile.	4.6	III. Usually foliated.	Basal, perfect.	Opaque. Greasy feel.	Infusible; reactions for sulphur.
2.	Malleable and sectile.	7.3	I. Also massive.	Uneven.	Opaque.	Reactions for sulphur; silver on charcoal.
2.	Brittle.	4.8	IV. Usually columnar or massive.	Prismatic.	Opaque.	Infusible; amethystine bead with borax; evolves chlorine with HCl.
2.	Brittle.	4.5	IV. Columnar to granular.	Pinacoidal, perfect.	Opaque.	Fuses very readily and gives antimony fumes and coating on charcoal.
2.	Brittle.	9.7	III. Also massive and foliated.	Basal, perfect.	Opaque. Tarnishes.	Fuses and volatilizes, leaving yellow coating.
2.5	Malleable and sectile.	7.3	I. Also massive.	Uneven.	Opaque.	Reactions for sulphur; silver on charcoal.
3.5	Brittle.	4.5	III. Usually massive.	Uneven.	Opaque. Tarnishes.	Fusible to magnetic mass, usually slightly magnetic before fusion.
4.	Brittle.	4.3	IV. Usually prismatic or columnar.	Prismatic, perfect.	Opaque.	Infusible; water in closed tube; amethystine bead with borax.
2.5	Brittle.	4.8	IV. Usually columnar or massive.	Prismatic.	Opaque.	Infusible; amethystine bead with borax; evolves chlorine with HCl.
3.	Brittle.	4.8	I. Usually massive.	Uneven.	Opaque.	Reactions for antimony; copper with soda on charcoal.
3.	Brittle to earthy.	6.25	IV. Usually massive or earthy.	Uneven or earthy.	Opaque.	Infusible; copper with soda on charcoal; green solution with nitric acid.
3.5	Brittle.	6.	III. Massive or botryoidal.	Basal, imperfect.	Opaque. Dark gray tarnish.	Volatilizes without fusing, giving gray fumes and coating on charcoal, and metallic sublimate in tube.
3.5	Brittle.	6.6	III. Lamellar or massive.	Basal, perfect.	Opaque.	Fuses readily on charcoal, giving copious white fumes and a white coating.
2.5	Brittle.	9.7	III. Also massive and foliated.	Basal, perfect.	Opaque. Tarnishes.	Fuses and volatilizes, leaving yellow coating on charcoal.

Analytical Key.			Species.	Composition.	Lustre.	Color.	Streak.
IV. Color Gray.—Continued.	2. Soft.—Continued.	Streak gray. Color dark gray to black on fresh surface.	**Stibnite** (13).	Sb_2S_3.	Metallic.	Light lead-gray.	Lead-gray.
			Chalcocite (9).	CuS.	Metallic.	Blackish lead-gray.	Blackish lead-gray.
			Galenite (6).	PbS.	Metallic.	Dark lead-gray.	Dark lead-gray.
	3. Hard.	Strongly magnetic. Malleable.	**Iron** (2).	Fe.	Metallic.	Iron-gray to black.	Iron-gray to black.
		Strongly magnetic. Brittle.	**Magnetite** (23).	Fe_3O_4.	Metallic.	Dark gray to iron-black.	Black.
		Streak grayish black. Color white to light gray on fresh surface.	**Marcasite** (11).	FeS_2.	Metallic.	Pale grayish yellow.	Grayish black.
			Arsenopyrite (11).	FeAsS.	Metallic.	Silver-white to steel-gray.	Grayish black.
			Cobaltite (10).	CoAsS.	Metallic.	Silver-white to steel-gray.	Grayish black.
			Smaltite (10).	$(CoFeNi)As_2$.	Metallic.	Tin-white to steel-gray.	Grayish black.
		Streak grayish or brownish black. Color yellowish or reddish on fresh surface.	**Pyrrhotite** (8).	Fe_7S_8.	Metallic.	Bronze-yellow.	Grayish black.
			Marcasite (11).	FeS_2.	Metallic.	Pale grayish yellow.	Grayish black.
			Niccolite (8).	NiAs.	Metallic.	Pale copper-red.	Brownish black.
		Streak brownish black. Color dark gray to black.	**Psilomelane** (32).	$MnO_2 + H_2O$.	Submetallic.	Iron-black to steel-gray.	Brownish black.
			Manganite (29).	$Mn_2O_3 + H_2O$.	Submetallic.	Iron-black to steel-gray.	Brownish black.
			Menaccanite (25).	$(FeTi)_2O_3$.	Submetallic.	Iron-black to steel-gray.	Brownish black.

H.	Tenacity.	G.	Form.	Cleavage.	Other Properties.	Confirmatory Chemical Tests.
2.5	Brittle.	4.5	IV. Columnar to massive.	Pinacoidal, perfect.	Opaque.	Fuses very readily, and gives antimony fumes and coating on charcoal.
2.75	Brittle.	5.7	IV. Crystals also massive.	Prismatic, indistinct.	Opaque. Dull black tarnish.	Fusible; blue flame with HCl; copper with soda on charcoal; sulphur reactions.
2.75	Brittle.	7.5	I. Cubic and massive.	Cubic, perfect.	Opaque.	Fusible, yielding metallic lead and sulphur fumes.
4.5	Malleable.	7.5	I. Usually compact.	Usually none.	Opaque, Strongly magnetic.	Infusible; soluble in HCl.
5.5	Brittle.	5.	I. Usually granular.	Octahedral.	Opaque. Strongly magnetic.	Infusible; soluble in HCl.
6.	Brittle.	4.8	IV. Often globular, etc.	Prismatic, perfect.	Opaque.	Magnetic residue on charcoal; sulphur sublimate in tube.
5.5	Brittle.	6.2	IV. Also massive.	Prismatic, distinct.	Opaque.	Arsenic coating and magnetic residue on charcoal; arsenic sublimate in tube.
5.5	Brittle.	6.2	I. Also massive.	Cubic, perfect.	Opaque.	Arsenic fumes and coating on charcoal; cobalt-blue with borax.
5.5	Brittle.	7.	I. Usually massive.	Cubic, in traces.	Opaque.	Arsenic coating and magnetic residue on charcoal; arsenic sublimate in tube.
4.5	Brittle.	4.5	III. Usually massive.	Uneven.	Opaque. Tarnishes.	Fusible to magnetic mass; usually slightly magnetic before fusion.
6.	Brittle.	4.8	IV. Often globular, etc.	Prismatic, perfect.	Opaque.	Magnetic residue on charcoal; sulphur sublimate in tube.
5.25	Brittle.	7.5	III. Usually massive.	Uneven.	Opaque.	Arsenic fumes and coating and magnetic globule on charcoal.
5.	Brittle.	4.2	Compact and often botryoidal.	Even or conchoidal.	Opaque.	Infusible; water in closed tube; evolves chlorine with HCl.
4.	Brittle.	4.3	IV. Usually prismatic or columnar.	Prismatic, perfect.	Opaque.	Infusible; water in closed tube; amethystine bead with borax.
5.	Brittle.	4.75	III. Often laminated.	None.	Opaque. Often slightly magnetic.	Infusible; reactions for iron.

Analytical Key.		Species.	Composition.	Lustre.	Color.	Streak.
		Chromite (23).	$FeCr_2O_4$.	Submetallic.	Iron-black to dark gray.	Grayish brown.
	Streak brown.	Franklinite (23).	$(FeZnMn)_3 O_4$.	Metallic.	Iron-black to dark gray.	Reddish brown.
	Streak red.	Hematite (25).	Fe_2O_3.	Metallic.	Steel-gray to iron-black.	Red.
	Strongly magnetic.	Magnetite(23).	Fe_3O_4.	Metallic.	Iron-black to dark gray.	Black.
		Emery (25).	$Al_2O_3+ Fe_3O_4$.	Metallic.	Iron-black to brown.	Black to brown.
	Streak grayish black. Color white, gray or grayish yellow, on fresh surface.	Marcasite (11).	FeS_2.	Metallic.	Pale grayish yellow.	Grayish black.
		Arsenopyrite (11).	FeAsS.	Metallic.	Silver-white to steel-gray.	Grayish black.
		Cobaltite (10).	CoAsS.	Metallic.	Silver-white to steel-gray	Grayish black.
		Smaltite (10).	$(CoFeNi) As_2$.	Metallic.	Tin-white to steel-gray.	Grayish black.
	Streak brownish black. Color dark gray to black.	Psilomelane (32).	$MnO_2 +H_2O$.	Submetallic.	Iron-black to steel-gray.	Brownish black.
		Menaccanite (25).	$(FeTi)_2O_3$.	Submetallic.	Iron-black to steel-gray.	Brownish black.
	Streak light gray or brown. G. about 7.	Cassiterite (26).	SnO_2.	Adamantine.	Gray to black.	Light gray or brown.
	Streak light gray or brown. G. about 4.	Rutile (26).	TiO_2.	Adamantine.	Brown to gray.	Light gray or brown.
	Streak dark reddish brown.	Franklinite (23).	$(FeZnMn)_3 O_4$.	Metallic.	Iron-black to dark gray.	Reddish brown.
	Streak red.	Hematite (25).	Fe_2O_3.	Metallic.	Steel-gray to iron-black.	Red.

IV. Color Gray—Continued.

3. Hard.— Continued.

4. Very Hard.

H.	Tenacity.	G.	Form.	Cleavage.	Other Properties.	Confirmatory Chemical Tests.
5.5	Brittle.	4.4	I. Usually massive.	None.	Opaque. Sometimes magnetic.	Infusible; green bead with borax.
5.5	Brittle.	5.	I. Octahedrons, also massive.	Indistinct.	Opaque. Slightly magnetic.	Infusible; reactions for manganese and zinc with borax and soda.
5.5	Brittle.	4.9	III. Scaly, massive or botryoidal.	Basal.	Opaque. Sometimes magnetic.	Infusible; magnetic on charcoal; soluble in HCl.
6.5	Brittle.	5.	I. Usually granular.	Octahedral.	Opaque. Strongly magnetic.	Infusible; soluble in HCl.
7to9	Brittle.	4.5	III. Usually granular.	None.	Opaque. Strongly magnetic.	Infusible and insoluble.
6.5	Brittle.	4.8	IV. Often globular, etc.	Prismatic, perfect.	Opaque.	Magnetic on charcoal; sulphur sublimate in closed tube.
6.	Brittle.	6.2	IV. Also massive.	Prismatic, distinct.	Opaque.	Arsenic coating and magnetic residue on charcoal; arsenic sublimate in tube.
6.	Brittle.	6.2	I. Also massive.	Cubic, perfect.	Opaque.	Arsenic fumes and coating on charcoal; cobalt-blue with borax.
6.	Brittle.	7.	I. Usually massive.	Cubic, in traces.	Opaque.	Arsenic coating and magnetic residue on charcoal; arsenic sublimate in tube.
6.	Brittle.	4.2	Compact and often botryoidal.	Even or conchoidal.	Opaque.	Infusible; water in closed tube; evolves chlorine with HCl.
6.	Brittle.	4.75	III. Often laminated.	None.	Opaque. Often slightly magnetic.	Infusible; reactions for iron.
6.5	Brittle.	6.8	II. Also massive and botryoidal.	Imperfect.	Opaque.	Infusible; metallic tin with soda on charcoal.
6.25	Brittle.	4.2	II. Prismatic and twin crystals.	Prismatic, distinct.	Opaque.	Infusible.
6.5	Brittle.	5.	I. Octahedrons, also massive.	Indistinct.	Opaque. Slightly magnetic.	Infusible; reactions for manganese and zinc with borax and soda.
6.5	Brittle.	4.9	III. Scaly, massive or botryoidal.	Basal.	Opaque. Sometimes magnetic.	Infusible; magnetic on charcoal; soluble in HCl.

Analytical Key.			Species.	Composition.	Lustre.	Color.	Streak.
V. Color White.	1. Very Soft.	Liquid.	**Mercury** (2).	Ag.	Metallic.	Tin-white.	White.
		Brittle.	**Bismuth** (3).	Bi.	Metallic.	Reddish white.	Gray.
		Malleable.	**Silver** (1).	Ag.	Metallic.	Silver-white.	White.
	2. Soft.	Brittle.	**Arsenic** (3).	As.	Metallic.	Tin-white.	White.
			Antimony (3).	Sb.	Metallic.	Tin-white.	White.
			Bismuth (3).	Bi.	Metallic.	Reddish white.	Gray.
		Malleable.	**Silver** (1).	Ag.	Metallic.	Silver-white	White.
	3. Hard.	Ortho-rhombic.	**Arsenopyrite** (11).	FeAsS.	Metallic.	Silver-white to steel-gray.	Grayish black.
		Isometric.	**Cobaltite** (10).	CoAsS.	Metallic.	Silver-white to steel-gray.	Grayish black.
			Smalite (10).	(CoFeNi)As$_2$.	Metallic.	Tin-white to steel-gray.	Grayish black.
	4. Very Hard.	Ortho-rhombic.	**Arsenopyrite** (11).	FeAsS.	Metallic.	Silver-white to steel-gray.	Grayish black.
		Isometric.	**Cobaltite** (10).	CoAsS.	Metallic.	Silver-white to steel-gray.	Grayish black.
			Smaltite (10).	(CoFeNi)As$_2$.	Metallic.	Tin-white to steel-gray.	Grayish black.

H.	Tenacity.	G .	Form.	Cleavage.	Other Properties.	Confirmatory Chemical Tests.
	Liquid.	13.5	Fluid globules.	None.	Opaque.	Volatile; dissolves in nitric acid.
2.	Brittle.	9.7	III. Also massive and foliated.	Basal, perfect.	Opaque. Tarnishes.	Fuses and volatilizes, leaving yellow coating on charcoal.
2.5	Malleable.	10.5	I. Grains, scales and threads.	None.	Opaque. Tarnishes black.	Infusible; soluble in nitric acid.
3.5	Brittle.	6.	III. Massive and botryoidal.	Basal, imperfect.	Opaque. Dark gray tarnish.	Volatilizes without fusing; gray fumes and coating on charcoal, and metallic sublimate in tube.
3.5	Brittle.	6.6	III. Lamellar or massive.	Basal, perfect.	Opaque.	Fuses readily on charcoal, giving copious white fumes and coating.
2.5	Brittle.	9.7	III. Also massive and foliated.	Basal, perfect.	Opaque. Tarnishes.	Fuses and volatilizes, leaving yellow, volatile coating on charcoal.
3.	Malleable.	10.5	I. Grains, scales and threads.	None.	Opaque. Tarnishes black.	Fusible; soluble in nitric acid.
5.5	Brittle.	6.2	IV. Crystals and massive.	Prismatic, distinct.	Opaque.	Arsenic coating and magnetic residue on charcoal; arsenic sublimate in tube.
5.5	Brittle.	6.2	I. Crystals and massive.	Cubic, perfect.	Opaque.	Arsenic fumes and coating on charcoal; blue bead with borax.
5.5	Brittle.	7.	I. Usually massive.	Cubic, in traces.	Opaque.	Like arsenopyrite; but blue bead with borax.
6.	Brittle.	6.2	IV. Crystals and massive.	Prismatic, distinct.	Opaque.	Arsenic coating and magnetic residue on charcoal; arsenic sublimate in tube.
5.5	Brittle.	6.2	I. Crystals and massive.	Cubic, perfect.	Opaque.	Arsenic fumes and coating on charcoal; blue bead with borax.
6.	Brittle.	7.	I. Usually massive.	Cubic, in traces.	Opaque.	Like arsenopyrite; but blue bead with borax.

I. Streak Red or Brown.

Analytical Key.		Species.	Composition.	Lustre.	Color.	Streak.
1. Very Soft.		Turgite (Red Ochre) (28). *Compare Hematite and red Kaolinite.*	$2Fe_2O_3 + H_2O$.	Dull.	Red.	Red.
		Realgar (12).	AsS.	Resinous.	Aurora-red.	Orange to aurora-red.
	Streak red.	Proustite (15).	Ag_3AsS_3.	Adamantine to dull.	Cochineal-red.	Cochineal-red.
		Pyrargyrite (15).	Ag_3SbS_3.	Adamantine to dull.	Black to cochineal-red.	Cochineal-red.
		Cinnabar (8). *G. varies with impurity from 3 to 9.*	HgS.	Adamantine to dull.	Cochineal-red to brown.	Scarlet.
	Streak yellow to brown.	Limonite (Yellow Ochre) (30).	$2Fe_2O_3 + 3H_2O$.	Dull.	Brown to yellow.	Light brown to yellow.
	Streak dark brown to black. G. 1 to 1.5	Asphaltum (84).	C,H,O, etc.	Resinous.	Brownish black to black.	Dark brown.
		Mineral Coal (84).	C,H,O. etc.	Resinous.	Brown to black.	Dark brown.
2. Soft.	Streak dark brown to black. G. 3 to 4.	Wad (32).	$2MnO_2 + H_2O$	Dull.	Brownish black.	Brownish black.
	Streak bright yellow to orange.	Zincite (21).	ZnO.	Sub-adamantine.	Red to orange.	Orange to yellow.
	Streak yellow to brown. Compact or fibrous.	Limonite (30). *Compare Gothite.*	$2Fe_2O_3 + 3H_2O$	Submetallic to dull.	Dark brown to yellow.	Yellow to brown.
		Turgite (28).	$2Fe_2O_3 + H_2O$	Submetallic to dull.	Red to black.	Red.
	Streak red. Become magnetic on charcoal.	Hematite (25).	Fe_2O_3.	Metallic to dull.	Red to black.	Red.
		Cuprite (20).	Cu_2O.	Adamantine to dull.	Red to brown.	Red to brown.
		Cinnabar (8).	HgS.	Adamantine to dull.	Cochineal-red to brown.	Scarlet.

H.	Tenacity.	G.	Form.	Cleavage.	Other Properties.	Confirmatory Chemical Tests.
1.	Earthy.	3.5	Compact.	Earthy.	Opaque.	Water in tube; · black and magnetic on charcoal.
1.75	Brittle.	3.5	V. Also massive.	Clinopinacoidal and basal.	Transparent to translucent.	Volatile and combustible, burning with a blue flame and arsenical odor.
2.	Brittle.	5.5	III. Also massive.	Conchoidal.	Translucent to opaque.	Reactions for arsenic and silver.
2.	Brittle.	5.8	III. Also massive.	Conchoidal.	Translucent to opaque.	Reactions for antimony and silver.
2.	Brittle to sectile.	9.	III. Usually massive.	Uneven.	Usually opaque.	Volatile; with soda in tube a sublimate of mercury.
1.	Earthy.	3.5	Compact.	Earthy.	Opaque.	Water in tube; black and magnetic on charcoal.
1.5	Brittle to flexible.	1.3	Amorphous.	Conchoidal.	Opaque.	Fuses and burns with flame.
1.5	Brittle.	1.25	Amorphous, often laminated.	Even or conchoidal.	Opaque.	Infusible; but readily ignited, burning with flame.
1.	Earthy.	3.8	Amorphous	Earthy.	Opaque.	Water in closed tube; amethystine bead with borax; evolves chlorine with HCl.
4.	Brittle.	5.5	III. Also massive.	Basal, perfect.	Usually opaque.	Infusible; soluble in acid; zinc coating with soda on charcoal.
3.	Brittle.	3.8	Compact, fibrous and botryoidal.	None.	Opaque.	Water in closed tube; black and magnetic on charcoal.
3.	Brittle.	3.7	Compact and botryoidal.	None.	Opaque.	Infusible; water in tube; magnetic on charcoal.
3.	Brittle.	4.9	III. Scaly, also compact and botryoidal.	Basal.	Opaque. Sometimes magnetic.	Like turgite; but no water in tube.
3.5	Brittle.	6.	I. Also capillary and massive.	Octahedral.	Transparent to opaque.	Colors flame green and fuses readily, yielding metallic copper; becomes magnetic when impure.
2.5	Brittle to sectile.	9.	III. Usually massive.	Uneven.	Usually opaque.	Volatile; with soda in closed tube yields sublimate of mercury; becomes magnetic when impure.

Analytical Key.			Species.	Composition.	Lustre.	Color.	Streak.
I. Streak Red or Brown.—Continued.	*2. Soft.—Continued.*	Streak red. Not magnetic on charcoal.	Proustite (15).	Ag_3AsS_3.	Adamantine to dull.	Cochineal-red.	Cochineal-red.
			Pyrargyrite (15).	Ag_3SbS_3.	Adamantine to dull.	Black to cochineal-red.	Cochineal-red.
			Cuprite (20).	Cu_2O.	Adamantine to dull.	Red to brown.	Red to brown.
			Cinnabar (8).	HgS.	Adamantine to dull.	Cochineal-red to brown.	Scarlet.
		Streak dull brown to black.	Wad (32).	$2MnO_2 + H_2O$.	Dull.	Brownish black.	Brownish black.
			Siderite (44).	$FeCO_3$.	Vitreous to dull.	Gray to dark brown.	Gray to brown.
			Sphalerite. (7).	ZnS.	Resinous.	Brown to black.	Yellow to brown.
	3. Hard.	Streak bright yellow to orange.	Zincite. (21).	ZnO.	Sub-adamantine.	Deep red to orange.	Orange-yellow.
		Streak yellow to brown. Compact or fibrous.	Limonite (30).	$2Fe_2O_3 + 3H_2O$.	Submetallic to dull.	Brown to black.	Yellow to brown.
			Göthite (29).	$Fe_2O_3 + H_2O$.	Adamantine to dull.	Brown to black.	Yellow to brown.
		Streak bright red.	Turgite (28).	$2Fe_2O_3 + H_2O$.	Submetallic to dull.	Reddish black.	Red.
			Hematite (25).	Fe_2O_3.	Metallic to dull.	Reddish black to steel-gray.	Red.
			Cuprite (20).	Cu_2O.	Submetallic to dull.	Red to brown.	Red to brown.
			Cinnabar (8).	HgS.	Adamantine to dull.	Cochineal-red to brown.	Scarlet.
		Streak reddish brown. H. 4.	Sphalerite (7).	ZnS.	Resinous.	Brown to black.	Yellow to brown.

H.	Tenacity.	G.	Form.	Cleavage.	Other Properties.	Confirmatory Chemical Tests.
2.5	Brittle.	3.7	III. Also massive.	Conchoidal.	Translucent to opaque.	Reactions for arsenic and silver.
2.5	Brittle.	4.9	III. Also massive.	Conchoidal.	Translucent to opaque.	Reactions for antimony and silver.
3.5	Brittle.	6.	I. Also capillary and massive.	Octahedral.	Translucent to opaque.	Colors flame green and fuses readily, yielding metallic copper.
2.5	Brittle to sectile.	9.	III. Usually massive.	Uneven.	Usually opaque.	Volatile; with soda in closed tube yields sublimate of mercury.
3.	Earthy.	3.8	Amorphous.	Earthy.	Opaque.	Water in closed tube; amethystine bead with borax; evolves chlorine with HCl.
3.5	Brittle.	3.8	III. Also compact.	Rhombohedral, perfect.	Translucent to opaque.	Infusible; blackens and becomes magnetic; effervesces in hot acid.
3.5	Brittle.	4.	I. Also massive.	Dodecahedral, perfect.	Translucent to opaque.	Reactions for sulphur; zinc coating with soda on charcoal.
4.5	Brittle.	5.5	III. Also massive.	Basal, perfect.	Usually opaque.	Infusible; soluble in acid; zinc coating with soda on charcoal.
5.25	Brittle.	4.	Compact, botryoidal, fibrous, etc.	Uneven, splintery, etc.	Opaque.	Water in closed tube; magnetic on charcoal.
5.25	Brittle.	4.2	IV. Usually massive, fibrous or botryoidal.	Prismatic, perfect.	Opaque.	Like limonite.
5.	Brittle.	3.7	Massive and botryoidal.	Uneven.	Opaque.	Water in tube; magnetic on charcoal.
5.5	Brittle.	4.9	III. Scaly, also massive or botryoidal.	Basal.	Opaque. Sometimes magnetic.	Infusible; magnetic on charcoal; soluble in HCl.
4.	Brittle.	6.	I. Also capillary and massive.	Octahedral.	Translucent to opaque.	Colors flame green and fuses readily, yielding metallic copper.
4.	Brittle to sectile.	9.	III. Usually massive.	Uneven.	Usually opaque.	Volatile; with soda in closed tube yields sublimate of mercury.
4.	Brittle.	4.	I. Usually massive.	Dodecahedral, perfect.	Translucent to opaque.	Reactions for sulphur and zinc; soluble in acid with effervescence.

Analytical Key.			Species.	Composition.	Lustre.	Color.	Streak.
I. Streak Red or Brown.—Continued.	3. Hard.—Continued.		Psilomelane (32).	MnO_2 $+H_2O$.	Submetallic.	Iron-black.	Brownish black.
		Streak dark reddish brown to black. H. above 5.	Menaccanite (25).	$(FeTi)_2O_3$.	Submetallic.	Iron-black.	Brownish black.
			Samarskite (35).	Complex columbate.	Submetallic, shining.	Black.	Reddish brown.
			Wolframite (41). Compare *Manganite and Columbite*.	$(FeMn)$ WO_4.	Submetallic.	Brownish black.	Brownish black.
		Streak grayish brown.	Siderite (44).	$FeCO_3$.	Vitreous.	Gray to dark brown.	Gray to brown.
			Sphalerite (7).	ZnS.	Resinous.	Brown to black.	Yellow to brown.
			Chromite (23).	$FeCr_2O_4$.	Submetallic.	Iron-black.	Grayish brown.
	4. Very Hard.	Streak bright red. G. 3.6	Turgite (28).	$2Fe_2O_3$ $+H_2O$.	Submetallic.	Reddish black.	Red.
		Streak pale red or brown. G. 2.6	Quartz (Ferruginous quartz, and Jasper) (33).	SiO_2 $+Fe_2O_3$.	Vitreous to dull.	Red to brown.	Light red to brown.
		Streak pale brown. G. above 4.	Rutile (26).	TiO_2.	Adamantine.	Brown.	Light brown.
			Cassiterite (26). Compare *Chromite*.	SnO_2.	Adamantine.	Brown to black.	Gray to light brown.
		Streak dark reddish brown to black.	Psilomelane (32).	MnO_2+ H_2O.	Submetallic.	Iron-black.	Brownish black.
			Menaccanite (25).	$(FeTi)_2O_3$.	Submetallic.	Iron-black.	Brownish black.
			Samarskite (35).	Complex columbate.	Submetallic, shining.	Black.	Reddish brown.
			Columbite (35).	$FeCb_2O_6$.	Submetallic.	Black.	Brownish black.

H.	Tenacity.	G.	Form.	Cleavage.	Other Properties.	Confirmatory Chemical Tests.
5.	Brittle.	4.2	Compact and botryoidal.	Even or conchoidal.	Opaque.	Infusible; water in closed tube; evolves chlorine with HCl.
5.	Brittle.	4.75	III. Often laminated.	None.	Opaque. Often slightly magnetic.	Infusible; reactions for iron.
5.5	Brittle.	5.7	IV. Also massive.	None.	Opaque.	Emerald-green bead with SPh; green bead with soda.
5.25	Brittle.	7.4	V. Also lamellar or massive.	Prismatic, perfect.	Opaque.	Fuses easily to a magnetic globule; reactions for manganese.
4.5	Brittle.	3.8	III. Also compact.	Rhombohedral, perfect.	Translucent to opaque.	Infusible; blackens and becomes magnetic; effervesces in hot acid.
4.	Brittle.	4.	I. Also massive.	Dodecahedral, perfect.	Translucent to opaque.	Reactions for sulphur; zinc coatings with soda on charcoal; effervesces in acid (H_2S).
5.5	Brittle.	4.4	I. Usually massive.	None.	Opaque. Sometimes magnetic.	Infusible; green bead with borax.
6.	Brittle.	3.7	Compact and botryoidal.	Uneven.	Opaque.	Infusible; magnetic on charcoal; soluble in HCl; water in closed tube.
7.	Brittle.	2.6	III. Often massive or compact.	None.	Translucent to opaque.	Infusible and insoluble.
6.25	Brittle.	4.2	II. Prismatic and twin crystals.	Prismatic, distinct.	Translucent to opaque.	Infusible and insoluble.
6.5	Brittle.	6.8	II. Also massive and botryoidal.	Imperfect.	Translucent to opaque.	Infusible; yields metallic tin with soda on charcoal.
6.	Brittle.	4.2	Compact and botryoidal.	Even or conchoidal.	Opaque.	Infusible; water in closed tube; evolves chlorine with HCl.
6.	Brittle.	4.75	III. Often laminated.	None.	Opaque. Often slightly magnetic.	Infusible; reactions for iron.
6.	Brittle.	5.7	IV. Also massive.	Uneven or conchoidal.	Opaque.	Emerald-green bead with SPh; green bead with soda.
6.	Brittle.	6.	IV. Usually in crystals.	Prismatic.	Opaque.	Infusible.

II. Streak Yellow.

Analytical Key.			Species.	Composition.	Lustre.	Color.	Streak.
I. Very Soft.	G. 2 to 2.5		**Sulphur** (4).	S.	Resinous.	Yellow.	Yellow.
			Kaolinite (Yellow Clay) (78).	$Al_2Si_2O_7$ $+2H_2O.$	Dull.	Yellow.	Yellow.
	G. 3.5 to 4.		**Limonite** (Yellow Ochre)(30).	$2Fe_2O_8$ $+3H_2O.$	Dull.	Yellow.	Yellow.
			Orpiment (13). Compare *Realgar*.	$As_2S_3.$	Resinous to pearly.	Yellow.	Yellow.
2. Soft.		Streak bright yellow to orange.	**Zincite** (21).	ZnO.	Sub-adamantine.	Red to orange.	Orange to yellow.
		Streak pale yellow. G. 6 to 7.	**Wulfenite** (41).	$PbMoO_4.$	Resinous to adamantine.	Yellow.	Yellowish white.
			Vanadinite (36). Compare *Pyromorphite and Mimetite*.	$3Pb_3V_2O_8$ $+PbCl_2.$	Resinous to dull.	Brown to yellow.	Yellowish.
		Streak yellow to brown. G. 3 to 4.	**Sphalerite** (7).	ZnS.	Resinous.	Yellow to brown.	Yellowish brown.
			Siderite (44).	$FeCO_3.$	Vitreous to dull.	Gray to dark brown.	Gray to brown.
			Limonite (30). Compare *Gothite*.	$2Fe_2O_8$ $+3H_2O.$	Submetallic to dull.	Dark brown to yellow.	Yellow to brown.
3. Hard.		Streak bright yellow to orange.	**Zincite** (21).	ZnO.	Sub-adamantine.	Red to orange.	Orange to yellow.
		Streak yellow to brown.	**Sphalerite** (7).	ZnS.	Resinous.	Yellow to brown.	Yellowish brown.
			Siderite (44).	$FeCO_3.$	Vitreous to dull.	Gray to dark brown.	Gray to brown.
			Limonite (30).	$2Fe_2O_8$ $+3H_2O.$	Submetallic to dull.	Brown to black.	Yellow to brown.
			Göthite (29).	Fe_2O_8 $+H_2O.$	Adamantine to dull.	Brown to black.	Yellow to brown.

H.	Tenacity.	G.	Form.	Cleavage.	Other Properties.	Confirmatory Chemical Tests.
2.	Brittle.	2.	IV. Often massive.	Imperfect.	Translucent. Electric by friction.	Burns with a blue flame and sulphurous odor.
1.	Plastic to earthy.	2.5	IV. Usually compact or clayey.	Earthy.	Opaque.	Infusible and insoluble; water in tube.
1.	Earthy.	3.7	Earthy.	None.	Opaque.	Black and magnetic on charcoal; water in closed tube.
2.	Sub-sectile.	3.5	IV. Usually foliated or massive.	Macropina-coidal, perfect.	Translucent.	Volatilizes; and gives reactions for arsenic and sulphur.
4.	Brittle.	5.5	III. Also massive.	Basal, perfect.	Usually opaque.	Infusible; zinc coating with soda on charcoal; soluble in acid.
3.	Brittle.	6.5	II. Very thin tabular crystals.	Octahedral, distinct.	Translucent.	Fuses readily; and yields lead with soda on charcoal.
3.	Brittle.	7.	III. Usually incrusting.	None.	Translucent to opaque.	Fuses readily; yields lead with soda on charcoal; and reacts for chlorine with copper oxide.
3.5	Brittle.	4.	I. Usually massive.	Dodecahedral, perfect.	Translucent to opaque.	Reactions for sulphur; zinc coating with soda on charcoal.
3.5	Brittle.	3.8	III. Also compact.	Rhombohedral, perfect.	Translucent to opaque.	Infusible; blackens and becomes magnetic; effervesces in hot acid.
3.	Brittle.	3.8	Compact, fibrous and botry'id'l.	None.	Opaque.	Water in closed tube; black and magnetic on charcoal.
4.5	Brittle.	5.5	III. Also massive.	Basal, perfect.	Usually opaque.	Infusible; zinc coating with soda on charcoal; soluble in acid.
4.	Brittle.	4.	I. Usually massive.	Dodecahedral, perfect.	Translucent to opaque.	Reactions for sulphur; zinc coating with soda on charcoal.
4.5	Brittle.	3.8	III. Also compact.	Rhombohedral, perfect.	Transluceut to opaque.	Infusible; blackens and becomes magnetic; efferveses in hot acid.
5.25	Brittle..	4.	Compact, botryoidal, fibrous, etc.	Uneven, splintery, etc.	Opaque.	Water in closed tube; magnetic on charcoal.
5.25	Brittle.	4.2	IV. Usually massive, fibrous or botryoidal.	Prismatic, perfect.	Opaque.	Like limonite.

Analytical Key.			Species.	Composition.	Lustre.	Color.	Streak.
II. Streak Yellow.—Continued.	**4. Very Hard.**	Compact.	**Quartz (Jasper)** (33).	SiO_2 $+Fe_2O_3$.	Nearly dull.	Brown to yellow.	Light yellow.
		Crystalline.	**Quartz (Ferruginous Quartz)** (33).	SiO_2 $+Fe_2O_3$.	Vitreous to nearly dull.	Brown to yellow.	Light yellow.
III. Streak Black.	**1 Very Soft.**		**Asphaltum** (84).	C,H,O,etc.	Resinous to dull.	Brownish black to black.	Dark brown.
		G. under 2.5	**Mineral Coal** (84).	C,H,O,etc.	Resinous to submetallic.	Brown to black.	Brown to black.
			Kaolinite (carbonaceous) (78).	$Al_2Si_2O_7$ $+2H_2O$.	Dull.	Black.	Black.
		G. 3.8	**Wad** (32).	MnO_2 $+H_2O$.	Dull.	Black.	Black.
	2. Soft.	G. under 2.	**Mineral Coal** (84).	C,H,O, etc.	Resinous to submetallic.	Brown to black.	Brown to black.
		G. nearly 4.	**Wad** (32).	MnO_2 $+H_2O$	Dull.	Black.	Black.
		G. over 6.	**Melaconite** (22).	CuO.	Metallic to dull.	Black.	Black.
	3. Hard.		**Psilomelane** (32).	MnO_2. $+H_2O$.	Submetallic.	Black.	Brownish black.
		G. under 6.	**Menaccanite** (25).	$(FeTi)_2O_3$.	Submetallic.	Black.	Brownish black.
			Samarskite (35).	Complex columbate.	Submetallic, shining.	Black.	Reddish brown.
		G. over 7.	**Wolframite** (41). Compare *Columbite.*	$(FeMn)$ WO_4.	Submetallic.	Brownish black.	Dark brown to black.
	4. Very Hard.	Crystalline. G. over 5.	**Columbite** (35). Compare *Psilomelane, Menaccanite and Samarskite.*	$FeCb_2O_6$.	Submetallic.	Black.	Dark red to black.

H.	Tenacity.	G.	Form.	Cleavage.	Other Properties.	Confirmatory Chemical Tests.
7.	Brittle.	2.6	III. Compact.	None.	Translucent to opaque.	Infusible and insoluble.
7.	Brittle.	2.6	III. Often massive.	None.	Translucent to opaque.	Infusible and insoluble.
1.5	Brittle to flexible.	1.3	Amorphous.	Conchoidal.	Opaque.	Fuses and burns with flame.
2.	Brittle.	1.5	Amorphous often laminated.	Even or conchoidal.	Opaque.	Infusible; burns.
1.	Plastic to earthy.	2.5	IV. Usually compact or earthy.	Earthy.	Opaque.	Infusible; water in tube.
1.	Earthy.	3.8	Amorphous.	IV. Earthy.	Opaque.	Water in closed tube; amethystine bead with borax; evolves chlorine with HCl.
2.5	Brittle.	1.5	Amorphous, often laminated.	Even or conchoidal.	Opaque.	Infusible; burns.
3.	Brittle to earthy.	3.8	Amorphous.	Uneven to earthy.	Opaque.	Water in closed tube; amethystine bead with borax; evolves chlorine with HCl.
3.	Brittle to earthy.	6.25	IV. Usually massive or earthy.	Uneven or earthy.	Opaque.	Infusible; copper with soda on charcoal; green solution with nitric acid.
5.	Brittle.	4.2	Compact or botryoidal.	Even or conchoidal.	Opaque.	Infusible; water in closed tube; evolves chlorine with HCl.
5.5	Brittle.	4.75	III. Often laminated.	None.	Opaque. Often slightly magnetic.	Infusible; reactions for iron.
5.5	Brittle.	5.7	IV. Also massive.	Uneven or conchoidal.	Opaque.	Emerald-green bead with SPh; green bead with soda.
5.25	Brittle.	7.4	V. Also lamellar or massive.	Prismatic, perfect.	Opaque.	Fuses easily to a magnetic globule; reactions for manganese.
6.	Brittle.	6.	IV. Usually in crystals.	Prismatic.	Opaque.	Infusible.

Analytical Key.			Species.	Composition.	Lustre.	Color.	Streak.
IV. Streak Green or Blue.	1. Very Soft.	Streak green.	Glauconite (76).	$(AlFeK)Si$ $O_8 + H_2O.$	Dull.	Dark green.	Lighter green.
			Prochlorite (81).	$(AlMgFe)_8$ $SiO_5 + H_2O.$	Pearly to dull.	Dark green.	Lighter green.
			Ripidolite (81).	$Al_2Mg_5Si_3O$ $O_{14} \cdot 4H_2$	Pearly.	Deep green.	Greenish.
		Streak blue.	Vivianite (37).	$Fe_3P_2O_8$ $+8H_2O.$	Vitreous to pearly or dull.	Blue to green.	Bluish.
	2. Soft.	Streak green. Not micaceous.	Atacamite (17).	$CuCl$ $+3H_2CuO_2.$	Adamantine to vitreous.	Green. -	Apple-green.
			Malachite (46). Compare *Chrysocolla.*	Cu_2CO_4 $+H_2O.$	Adamantine, silky or dull.	Green.	Paler green.
		Streak green. Micaceous.	Lepidomelane (56). Compare *Ripidolite.*	$(AlFeK)_2$ $SiO_4.$	Adamantine to pearly.	Black.	Grayish green.
		Streak blue.	Azurite (46).	$Cu_3C_2O_7$ $+H_2O.$	Vitreous to dull.	Blue.	Paler blue.
			Chrysocolla (68).	$CuSiO_3$ $+2H_2O.$	Vitreous to dull.	Bluish green.	Bluish.
	3. Hard.	Streak greenish.	Malachite (46).	Cu_2CO_4 $+H_2O.$	Adamantine, silky or dull.	Green.	Paler green.
			Pyroxene (Augite) (47).	$(CaMgAl$ $Fe)SiO_3.$	Vitreous.	Greenish black.	Grayish green.
			Amphibole (49) (Hornblende).	$(CaMgAl$ $Fe)SiO_3.$	Vitreous.	Greenish black.	Grayish green.
		Streak blue.	Azurite (46).	$Cu_3C_2O_7$ $+H_2O.$	Vitreous to dull.	Blue.	Paler blue.
			Lapis-Lazuli (57).	$(CaNaAl)_2$ $SiO_4.$	Vitreous	Blue.	Blue.
	4. Very Hard.	Streak bluish or greenish.	Turquois (38). Compare *Lapis-Lazuli, Pyroxene and Amphibole.*	$Al_2P_2O_8$ $+5H_2O.$	Waxy.	Blue.	Bluish to greenish white.

H.	Tenacity.	G.	Form.	Cleavage.	Other Properties.	Confirmatory Chemical Tests.
1.	Granular and earthy.	2.3	Amorphous.	Earthy.	Opaque.	Water in closed tube.
1.5	Somewhat sectile.	2.8	III. Usually finely foliated.	Basal, perfect.	Translucent to opaque.	Fuses with difficulty; water in closed tube.
2.25	Flexible and sectile.	2.7	V. Foliated or micaceous.	Basal, perfect.	Emerald-green by trans. light.	Like prochlorite.
2.	Sectile to earthy.	2.6	V. Often massive or earthy.	Clinopinacoidal, perfect.	Translucent to opaque.	Fuses easily to magnetic globule; water in closed tube; soluble in acid.
3.25	Brittle.	3.75	IV. Also massive.	Brachypinacoidal, perfect.	Transparent to opaque.	Fuses with a blue flame; yields metallic copper· water in closed tube.
3.75	Brittle.	3.9	V. Often botryoidal, fibrous o earthy.	Basal, perfect.	Translucent to opaque.	Fuses and yields copper; water in closed tube; effervesces with acid.
3.	Sectile to brittle.	3.	III. Foliated or micaceous.	Basal, perfect.	Translucent to opaque.	Fuses to a magnetic globule; decomposed by HCl.
4.	Brittle.	3.7	V. Also massive.	Clinodome, perfect.	Translucent to opaque.	Like malachite.
3.	Brittle to sectile.	2.2	Amorphous or botryoidal.	Conchoidal or uneven.	Translucent to opaque.	Infusible; water in closed tube; copper on charcoal; decomposed by HCl.
3.75	Brittle.	3.9	V. Often botryoidal, fibrous or earthy.	Basal, perfect.	Translucent to opaque.	Fuses and yields copper; water in closed tube; effervesces with acid.
5.5	Brittle.	3.4	V. Massive or fibrous.	Prismatic, perfect.	Translucent to opaque.	Fusible; insoluble.
5.5	Brittle.	3.2	V. Massive or fibrous.	Prismatic, perfect.	Translucent to opaque.	Like pyroxene.
4.	Brittle.	3.7	V. Also massive.	Clinodome, perfect.	Translucent to opaque.	Fuses and yields copper; water in closed tube; effervesces with acid.
5.5	Brittle.	2.4	I. Usually compact.	None.	Opaque.	Fusible; gelatinizes with HCl.
6.	Brittle.	2.7	Amorphous.	None.	Opaque.	Infusible; water in closed tube; soluble in acid.

Analytical Key.		Species.	Composition.	Lustre.	Color.	Streak.
	,	G. 5.5 Waxy.	Cerargyrite(16).	AgCl.	Resinous to dull.	Gray.	Gray, shining.
		G. 1. Like resin.	Amber and Copal (84).	C,H,O, etc.	Resinous.	Yellow and brown to white.	White.
		Micaceous. Foliæ highly elastic.	Muscovite (56). Compare *Hydromica.*	AlKSiO$_4$.	Pearly.	White, brown, green, etc.	White.
		Micaceous. Foliæ somewhat elastic.	Ripidolite (81). Compare *Prochlorite.*	Mg$_5$Al$_2$Si$_8$ O$_{14}$+4H$_2$O.	Pearly.	Deep green.	White.
V. Streak White or Gray.	1. Very Soft.	Greasy feel. Foliated or compact.	Talc (75).	MgSiO +H$_2$O.	Pearly to dull.	Green to gray, etc.	White.
			Pyrophyllite (75). Compare *Gypsum.*	Al$_2$Si$_8$O$_9$ +H$_2$O.	Dull to pearly.	White, gray, green, etc.	White.
		Compact. Floats on water when dry.	Sepiolite (76).	Mg$_2$Si$_8$O$_8$ +2H$_2$O.	Dull.	White to gray.	White.
		Compact, earthy or clayey.	Kaolinite (Clay) (78).	Al$_2$Si$_2$O$_7$ +2H$_2$O.	Dull to pearly.	White, gray, yellow, red, blue, green, etc.	Like color.
			Calcite (Chalk) (44).	CaCO$_3$.	Dull.	White.	White.
			Opal (Tripolite and Siliceous Tufa) (34).	SiO$_2$. +H$_2$O.	Dull.	White to gray.	White to gray.
		Blue or green.	Prochlorite(81).	(AlMgFe)$_3$. SiO$_5$+H$_2$O	Pearly to dull.	Dark green.	Gray.
		Soluble in water, having taste.	Sassolite (39).	H$_6$B$_2$O$_6$.	Pearly.	White.	White.
			Borax (40). Compare *Halite.*	Na$_2$B$_4$O$_7$ +10H$_2$O.	Vitreous to resinous.	White.	White.
		Very light, fibrous nodules.	Ulexite (40).	NaCaB$_2$O$_9$ +5H$_2$O.	Silky to dull.	White.	White.
		Yields much water in closed tube.	Gypsum (43).	CaSO$_4$ +2H$_2$O.	Pearly to silky and dull.	White, gray, brown, etc.	White.

H.	Tenacity.	G.	Form.	Cleavage.	Other Properties.	Confirmatory Chemical Tests.
1.5	Waxy and sectile.	5.5	I. Usually compact.	None.	Transparent to opaque.	Fuses; silver on charcoal; reacts for chlorine with CuO and SPh.
2.25	Brittle.	1.	Amorphous.	None.	Transparent to translucent.	Fuses readily and burns with a yellow flame.
2.	Sectile and elastic.	2.8	V. Usually foliated or lamellar.	Basal, perfect.	Transparent. to translucent.	Infusible and insoluble.
2.	Flexible and sectile.	2.7	V. Foliated or mica-ceous.	Basal, perfect.	Emerald-green by trans. light.	Fuses with difficulty; water in closed tube.
1.	Sectile and inelastic.	2.6	IV. Usually foliated, sometimes compact	Basal, perfect	Greasy feel. Translucent.	Infusible and insoluble; little water in closed tube.
1.5	Somewhat sectile.	2.8	IV. Usually compact, sometimes foliated.	Basal, perfect.	Greasy feel. Translucent to opaque.	Fuses with difficulty; yields water; blue color with cobalt solution.
2.25	Somewhat sectile.	Very light.	Compact.	None.	Opaque.	Infusible; water in closed tube; pink color with cobalt solution; gelatinizes with HCl.
1.5	Plastic or earthy.	2.5	IV. Usually compact or clayey.	Basal, perfect. Finely scaly.	Translucent to opaque.	Infusible and insoluble; water in closed tube; blue color with cobalt solution.
1.	Earthy.	2.6	III. Compact.	Earthy.	Opaque.	Infusible; soluble with efferves-cence; alkaline reaction after heating.
1.	Earthy.	1.7	Amorphous.	Earthy.	Opaque.	Infusible; water in tube.
1.5	Somewhat sectile.	2.8	III. Usually finely foli-ated.	Basal, perfect.	Transluceut to opaque.	Fuses with difficulty; water in closed tube.
1.	Brittle.	2.4	VI. Scaly.	Basal, perfect.	Translucent. Taste acidulous.	Fuses to a clear glass, coloring flame green; soluble in water; water in closed tube.
2.25	Brittle.	2.3	V. Usually crystalized.	Perfect.	Translucent to opaque. Taste alkaline.	Puffs up and fuses to clear glass, coloring flame yellow; soluble in water; water in closed tube.
1.	Sectile.	1.65	Nodules.	Finely fibrous.	Opaque.	Fuses readily to a clear glass, col-oring flame yellow; with H_2SO_4 the flame is green.
2.	Sectile to brittle.	2.3	V. Usually foliated, massive or fibrous.	Clinopina-coidal, perfect.	Transparent to opaque.	Fuses; water in closed tnbe; sul-phur reaction with soda on char-coal.

Analytical Key.				Species.	Composition.	Lustre.	Color.	Streak.
				Pyromorphite (36).	$3Pb_3P_2O_8$ $+PbCl_2$.	Resinous.	Green to yellow and brown.	White.
			Green, yellow or brown.	Mimetite (36).	$3Pb_3As_2O_8$ $+PbCl_2$.	Resinous.	Pale yellow to brown.	White.
				Vanadinite (36).	$3Pb_2V_2O_8$ $+PbCl_2$.	Resinous.	Yellow to brown.	Yellowish white.
		G. above 6.		Wulfenite (41).	$PbMoO_4$.	Resinous.	Yellow to green and brown.	White.
			White.	Anglesite (42).	$PbSO_4$.	Adamantine to vitrebus.	White.	White.
V. Streak White or Gray.—Continued.	*2. Soft.*			Cerussite (45).	$PbCO_3$.	Adamantine to vitreous.	White to gray.	White.
		G. above 3.5.		Rhodochrosite (44).	$MnCO_3$.	Vitreous.	Rose-red to brown.	White.
				Siderite (44).	$FeCO_3$.	Vitreous.	Gray to brown.	White to gray.
			Red, brown or gray.	Smithsonite (44).	$ZnCO_3$.	Vitreous to dull.	Grayish, greenish or brownish white.	White to gray.
				Sphalerite (7).	ZnS.	Resinous.	Pale yellow to brown.	White to gray.
		G. below 6.		Calamine (69).	Zn_2SiO_4 $+H_2O$.	Vitreous to dull.	Gray, yellow to brown.	White to gray.
				Barite (42).	$BaSO_4$.	Vitreous.	White to bluish or brownish.	White.
			White or bluish or greenish.	Celestite (42).	$SrSO_4$.	Vitreous.	White to bluish.	White.
				Witherite (45).	$BaCO_3$.	Vitreous.	White	White.
				Strontianite (45). Compare *Smithsonite.*	$SrCO_3$.	Vitreous.	White to greenish.	White.

H.	Tenacity.	G.	Form.	Cleavage.	Other Properties.	Confirmatory Chemical Tests.
3.75	Brittle.	6.8	III. Often botryoidal or massive.	Traces.	Translucent.	Readily fusible; lead with soda on charcoal; chlorine with CuO and SPh; white sublimate in tube.
3.5	Brittle.	7.2	III. Also massive.	Imperfect.	Translucent.	Like pyromorphite, but gives arsenical odor on charcoal.
3.	Brittle.	7.	III. Usually in incrustations.	None.	Opaque.	Like pyromorphite.
3.	Brittle.	6.7	II. Very thin, square, tabular crystals.	Octahedral.	Translucent.	Fuses readily; lead on charcoal with soda.
3.	Brittle.	6.25	IV. Aso massive and stalactitic.	Prismatic, interrupted.	Transparent to opaque.	Fuses readily; lead with soda on charcoal; coin test for sulphur.
3.5	Brittle.	6.5	IV. Also massive and stalactitic.	Prismatic, imperfect.	Transparent to opaque.	Fuses readily; lead with soda on charcoal; effervesces with acid.
4.	Brittle.	3.6	III. Also massive and botryoidal.	Rhombohedral, perfect.	Translucent.	Infusible; amethystine bead with borax; effervesces with acid.
4.	Brittle.	3.8	III. Also compact and concretionary.	Rhombohedral, perfect.	Translucent to opaque.	Infusible; blackens and becomes magnetic; effervesces with acid.
2 to 4.	Brittle to friable.	4.	III. Compact to earthy.	None.	Opaque.	Infusible; effervesces with acid; zinc coating on charcoal.
3.5	Brittle.	4.	I. Also massive.	Dodecahedral, perfect.	Translucent to opaque.	Infusible; zinc coating with soda on charcoal; evolves H_2S with HCl.
2 to 4.	Brittle to friable.	3.5	IV. Compact to earthy.	None.	Opaque.	Infusible; yields water; green color with cobalt solution; gelatinizes with HCl.
3.	Brittle.	4.5	IV. Also massive and stalactitic.	Basal and prismatic, perfect.	Transparent to opaque.	Fusible; colors flame yellowish green; sulphur reaction with soda on silver.
3.25	Brittle.	4.	IV. Also massive, fibrous, globular, etc.	Basal and prismatic, perfect.	Transparent to opaque.	Like barite; but colors flame red.
3.5	Brittle.	4.3	IV. Also globular, etc.	Prismatic, distinct.	Translucent to opaque.	Fusible; colors flame yellowish green; effervesces with acid.
3.75	Brittle.	3.7	IV. Also globular, etc.	Prismatic, perfect.	Transparent to opaque.	Infusible; colors flame red; effervesces with acid.

Analytical Key.			Species.	Composition.	Lustre.	Color.	Streak.
		Foliæ elastic. Black, even in thin foliæ.	Biotite (56).	$(K_2MgFe$ $Al)_2SiO_4.$	Pearly, splendent.	Black to deep green and brown.	White.
			Lepidomelane (56).	$(K_2FeAl)_2$ $SiO_4.$	Adamantine to pearly.	Black.	Grayish green.
		Foliæ somewhat elastic. Deep green.	Ripidolite (81).	$Al_2Mg_5Si_8$ $O_{14}+4H_2O.$	Pearly.	Deep green.	Greenish.
	Micaceous.		Phlogopite (56).	$(K_2MgAl)_2$ $SiO_4.$	Pearly to submetallic.	Yellowish brown.	White.
		Foliæ elastic. Not belonging to the foregoing.	Muscovite (56).	$(K_2Al)_2$ $SiO_4.$	Pearly.	White, brown, pale green, etc.	White.
			Lepidolite (56).	$(Li_2K_2Al)_2$ $SiO_4.$	Pearly.	Rose-red and violet to white.	White.
			Hydromica(80).	$(K_2Al)_2$ $SiO_4)$ $+H_2O.$	Pearly.	White to greenish.	White.
		Foliæ brittle.	Margarite (82).	$(CaFeAl)_3$ $SiO_5+H_2O.$	Pearly.	Reddish white to gray.	White.
		Dilute HCl.	Calcite (44).	$CaCO_3.$	Vitreous to dull.	White, and all colors when impure.	White to gray.
	Effervesce with		Aragonite (45).	$CaCO_3.$	Vitreous to dull.	White to gray, etc.	White.
		Strong HCl.	Dolomite (44).	$(CaMg)$ $CO_3.$	Vitreous to dull.	White to gray, etc.	White.
			Magnesite (44).	$MgCO_3.$	Vitreous to dull.	White to gray, etc.	White.
		Saline taste.	Halite (16).	$NaCl.$	Vitreous.	White to gray, brown, etc.	White.
		Acicular or fibrous.	Wavellite (38).	$Al_3P_4O_{19}$ $+12H_2O.$	Vitreous.	White to green, brown, etc.	White.
			Serpentine (Chrysotile) (77). Compare *Asbestus.*	$Mg_2SiO_4.$ $+H_2O.$	Silky.	Green to yellow.	White.

V. Streak White or Gray.—Continued.
2. Soft.—Continued.
G. below 3.5.

H.	Tenacity.	G.	Form.	Cleavage.	Other Properties.	Confirmatory Chemical Tests.
2.75	Elastic and sectile.	2.9	III. Tabular or foliated.	Basal, very perfect.	Transparent to opaque.	Fuses with difficulty; deocmposed by H_2SO_4; reactions for iron.
3.	Sectile to brittle.	3.	III. Tabular or foliated.	Basal, perfect.	Translucent to opaque.	Fuses to magnetic globule; decomposed by HCl.
2.5	Flexible and sectile.	2.7	V. Tabular or foliated.	Basal, perfect.	Transparent. Emerald-green by trans. light.	Fuses with difficulty; much water in closed tube.
2.75	Elastic and sectile.	2.8	IV. Tabular or foliated.	Basal, very perfect.	Transparent to translucent.	Fuses with difficulty; decomposed by H_2SO_4.
2.5	Elastic and sectile.	2.8	V. Tabular or foliated.	Basal, very perfect.	Transparent to translucent.	Infusible and insoluble; water in closed tube.
3.	Elastic and sectile.	2.9	IV. Tabular or foliated.	Basal, very perfect.	Translucent.	Fuses with intumescence, coloring the flame purplish red.
2.5	Elastic and sectile.	2.75	Tabular or foliated.	Basal, perfect.	Translucent. Greasy feel.	Infusible; water in closed tube.
4.	Laminæ stiff and brittle.	3.	IV. Usually lamellar or foliated.	Basal, perfect.	Translucent.	Infusible; whitens; water in closed tube.
3.	Brittle.	2.7	III. Rhombohedrons, also massive, compact, etc.	Rhombohedral, perfect.	Strong double refraction.	Infusible; glows; alkaline reaction after heating; effervesces in cold dilute acid.
3.75	Brittle.	2.9	IV. Prismatic, also stalactitic, etc.	Prismatic, imperfect.	Transparent to translucent.	Like calcite, but whitens and falls to pieces.
3.5	Brittle.	2.85	III. Rhombohedrons, also massive and compact.	Rhombohedral, imperfect.	Translucent to opaque.	Infusible; alkaline reaction after heating; does not effervesce freely in cold, dilute HCl.
3.5	Brittle.	3.	III. Usually compact.	Rhombohedral, perfect.	Transparent to opaque.	Like dolomite.
2.5	Brittle.	2.2	I. Cubes, also massive.	Cubic, perfect.	Transparent to translucent. Saline taste.	Fuses, coloring flame deep yellow; soluble in water.
3.75	Brittle.	2.3	IV. Usually in radiate, globular masses.	Prismatic, perfect to fibrous.	Translucent.	Swells up, colors flame green, but is infusible; blue color with cobalt solution; much water in tube.
3.	Flexible and sectile.	2.2	Fibrous veins.	Delicately fibrous.	Translucent.	Infusible; water in closed tube.

Analytical Key.			Species.	Composition.	Lustre.	Color.	Streak.
V. Streak White or Gray.—Continued. / *2. Soft.—Continued.* / *G. below 3.5.*	*Distinctly crystalline or cleavable.*	Octahedral cleavage. Cubic crystals.	Fluorite (18).	CaF_2.	Vitreous.	White, green, violet to blue, yellow, etc.	White.
		Only one perfect cleavage, tabular or foliated.	Brucite (31).	$MgO + H_2O$.	Pearly.	White to greenish.	White.
			Stilbite (74).	$(CaNa_2)Al_2 Si_6O_{16} + 6H_2O$.	Pearly.	White to yellow and red.	White.
			Heulandite(74).	$(CaNa_2)Al_2 Si_6O_{16} + 5H_2O$.	Pearly.	White to red, gray and brown.	White.
		Not belonging to the foregoing.	Cryolite (19).	$Na_6Al_2F_{12}$.	Vitreous.	White to brown.	White.
			Laumontite (67).	$CaAl_2Si_4O_{12} + 4H_2O$.	Vitreous.	White to red.	White.
			Anhydrite (42).	$CaSO_4$.	Vitreous to pearly.	White to bluish or reddish.	White.
	Compact or imperfectly crystalline.	Yields little or no water in closed tube.	Cryolite (19).	$Na_6Al_2F_{12}$.	Vitreous.	White to brown.	White.
			Anhydrite (42).	$CaSO_4$.	Vitreous to pearly.	White to bluish or reddish.	White.
		Bluish green.	Chrysocolla (68).	$CuSiO_3 + 2H_2O$.	Vitreous to dull.	Bluish green to blue.	Whitish.
		White.	Gibbsite (31).	$Al_2O_3 + 3H_2O$.	Vitreous to pearly and dull.	White to gray.	White.
			Sepiolite (76).	$Mg_2Si_3O_8\ 2H_2O$.	Dull.	White.	White.
		Yields water in closed tube.	Serpentine (77).	$Mg_3Si_2O_7 + 2H_2O$.	Resinous, waxy, greasy.	Green, yellow, brown, etc.	White.
			Dewylite (77).	$H_2Mg_2Si_3O_9 + 4H_2O$.	Resinous to greasy.	Yellow to gray and greenish.	White.
			Pinite (79).	$K_2Al_4Si_5O_{17} + 3H_2O$.	Waxy to dull.	Gray to green.	White.

H.	Tenacity.	G.	Form.	Cleavage.	Other Properties.	Confirmatory Chemical Tests.
4.	Brittle.	3.2	I. Cubes, also massive.	Octahedral, perfect.	Phosphorescent when heated.	Decrepitates, fuses and gives alkaline reaction; reaction for fluorine with $HKSO_4$.
2.5	Flexible and sectile.	2.4	III. Usually foliated or fibrous.	Basal, perfect.	Translucent.	Infusible; alkaline after heating; red color with cobalt solution; water in closed tube; soluble.
3.75	Brittle.	2.1	IV. Usually in sheaf-like aggregates.	Brachy-pinacoidal, perfect.	Transparent to translucent.	Fuses with intumescence; yields water; decomposed by HCl without gelatinizing.
3.75	Brittle.	2.2	V. Also globular and massive.	Clinopinacoidal, perfect.	Transparent to translucent.	Like stilbite.
2.5	Brittle.	3.	VI. Usually massive.	In three directions.	Translucent.	Fuses very easily, coloring flame yellow; fluorine reaction.
4.	Brittle.	2.3	V. Also columnar and massive.	Perfect.	Opaque and pulverulent on exposure.	Fuses with intumescence; yields water; gelatinizes with acid.
3.25	Brittle.	2.9	IV. Also fibrous, lamellar and massive.	Pinacoidal and basal, perfect.	Transparent to opaque.	Fusible; alkaline reaction; sulphur reaction with soda and silver.
2.5	Brittle.	3.	VI. Usually massive.	Imperfect.	Translucent.	Fuses very easily, coloring flame yellow; fluorine reaction.
3.25	Brittle.	2.9	IV. Often compact.	None apparent.	Translucent to opaque.	Fusible; alkaline reaction; sulphur reaction with soda and silver.
3.	Brittle to sectile.	2.2	Compact, sometimes botryoidal.	None.	Translucent to opaque.	Infusible; colors flame green; yields water; copper with soda on charcoal.
3.	Tough.	2.4	V. Usually stalactitic or incrusting.	Basal, perfect.	Translucent. Argillaceous odor.	Infusible; yields water; blue color with cobalt solution.
2.5	Somewhat sectile.	Very light.	Compact.	None.	Opaque. Absorbs water.	Infusible; water in closed tube; pink color with cobalt solution; gelatinizes with HCl.
3.5	Brittle.	2.6	Compact or finely granular.	None.	Translucent. Smooth feel.	Infusible; yields water.
2.5	Brittle.	2.2	Compact.	None.	Translucent.	Infusible; yields water.
3.	Brittle.	2.7	Compact.	None.	Translucent to opaque.	Fusible; yields water.

V. Streak White or Gray.—Continued.

3. Hard.

Analytical Key.			Species.	Composition.	Lustre.	Color.	Streak.
G. 4 or above 4 · *Distinct crystals also rare.*		Small square crystals.	Brookite (27).	TiO_2.	Adamantine.	Brown to yellowish and reddish.	White to gray.
		Gelatinizes with acid. Compare *Calamine.*	Willemite (51).	Zn_2SiO_4.	Vitreous to resinous, weak.	Yellowish, greenish and brownish.	White.
		Effervesces with hot acid (CO_2).	Smithsonite (44).	$ZnCO_3$.	Vitreous to dull.	White to grayish, greenish or brownish.	White.
		Effervesces with hot acid (H_2S). Compare *Allanite.*	Sphalerite (7).	ZnS.	Resinous.	Pale yellow to brown.	White to gray.
G. below 4 · *Very dark colored—greenish and brownish to black.*		Cleavage prismatic and nearly rectangular.	Enstatite (47).	$MgSiO_3$.	Vitreous to pearly.	Grayish and greenish to brown.	White to gray.
			Hypersthene (47).	$(MgFe)SiO_3$.	Pearly to metalloidal.	Dark brownish green to black.	Gray.
			Pyroxene (Augite) (47).	$(CaMgAl Fe)SiO_3$.	Vitreous.	Dark green to black.	Greenish gray.
		Cleavage prismatic and distinctly oblique.	Amphibole (Hornblende) (49).	$(CaMgAl Fe)SiO_3$.	Vitreous.	Dark green to black.	Greenish gray.
			Titanite (65).	$CaTiSiO_5$.	Adamantine to resinous.	Dark brown to black.	White to gray.
		Cleavage basal, perfect.	Chloritoid (82).	$(AlFeMg)_4 SiO_6 + H_2O$.	Pearly.	Dark gray and greenish to black.	Grayish to greenish.
		No distinct cleavage.	Allanite (55)	Complex silicate.	Submetallic to resinous.	Pitch-brown to black.	Gray.
		Blue.	Lazulite (38).	$MgAlP_2O_9 + H_2O$.	Vitreous.	Blue.	White.
			Sodalite (57).	$3Na_2Al_2Si_2 O_8 + 2NaCl$.	Vitreous to greasy.	Blue, etc.	White.
		Red to brown. Weathers black.	Rhodonite (48).	$MnSiO_3$.	Vitreous.	Red and brown to gray.	White.
		Effervesces with hot acid.	Magnesite (44). Compare *Cancrinite.*	$MgCO_3$.	Vitreous to dull.	White to gray, etc.	White.

H.	Tenacity.	G.	Form.	Cleavage.	Other Properties.	Confirmatory Chemical Tests.
5.5	Brittle.	4.2	IV. Rarely massive.	Imperfect.	Translucent.	Infusible and insoluble.
5.5	Brittle.	4.	III. Usually massive.	Imperfect.	Translucent.	Fuses with difficulty; gelatinizes with HCl.
5.	Brittle.	4.25	III. Often botryoidal or earthy.	Rhombohe-dral, perfect.	Translucent to opaque.	Infusible; effervesces with HCl; zinc coating with soda on char-coal.
4.	Brittle.	4.	I. Also massive.	Dodecahe-dral, perfect.	Translucent to opaque.	Infusible; zinc coating with soda on charcoal; evolves H_2S with HCl.
5.5	Brittle.	3.2	IV. Also massive and lamellar.	Prismatic, distinct.	Translucent to opaque.	Infusible and insoluble.
5.5	Brittle.	3.4	IV. Also massive and lamellar.	Prismatic, distinct.	Translucent to opaque.	Fusible to a black magnetic mass on charcoal.
5.	Brittle.	3.4	V. Often massive and granular.	Prismatic, perfect.	Translucent to opaque.	Fusible; insoluble.
5.	Brittle.	3.2	V. Often bladed or massive.	Prismatic, perfect.	Translucent to opaque.	Fusible; insoluble.
5.	Brittle.	3.5	V. Squarish, wedge-shap-ed crystals.	Prismatic, perfect.	Translucent to opaque.	Fuses with intumescence.
5.75	Brittle.	3.5	V. or VI. Coarsely foli-ated or scaly.	Basal, perfect.	Translucent to opaque.	Infusible; yields water.
5.75	Brittle.	3.6	V. Also massive and granular.	In traces.	Opaque.	Fuses with intumescence to a mag-netic mass.
5.	Brittle.	3.	V. Octahe-drons, also massive.	Imperfect.	Opaque.	Whitens and yields water; blue color is restored by cobalt solu-tion.
5.5	Brittle.	2.3	I. Dodecahe-drons and massive.	Dodecahe-dral, distinct.	Translucent.	Fuses with intumescence; gelatin-izes with HCl.
5.5	Tough.	3.5	VI. Com-pact or massive.	Indistinct.	Translucent to opaque.	Fusible; amethystine bead with borax.
4.5	Brittle.	3.	III. Usually compact.	Rhombohe-dral, perfect.	Transparent to opaque.	Infusible; alkaline reaction after heating; does not effervesce freely in cold dilute acid.

V. Streak White or Gray—Continued.
3. Hard.—Continued.
G. below 4.—Continued.

Analytical Key.	Species.	Composition.	Lustre.	Color.	Streak.
Crystallizes in very slender prisms, or in bladed, columnar or fibrous forms. — Little or no water in closed tube.	Wollastonite (47).	$CaSiO_3$.	Vitreous.	White to reddish.	White.
	Pyroxene (47).	$(CaMgFe)SiO_3$.	Vitreous.	Green to white.	White.
	Amphibole (Actinolite, Asbestus) (49).	$(CaMgFe)SiO_3$.	Vitreous to silky.	Green.	White.
	Amphibole (Tremolite, Asbestus) (49). Compare *Cyanite*.	$(CaMg)SiO_3$.	Vitreous to silky.	White.	White.
Much water in closed tube.	Pectolite (67).	$Na_2Ca_4Si_6O_{17}+H_2O$.	Silky to vitreous.	White.	White.
	Thomsonite (71).	$2(CaNa_2)Al_2Si_2O_8+5H_2O$.	Vitreous to pearly.	White.	White.
	Natrolite (71).	$Na_2Al_2Si_3O_{10}+2H_2O$.	Vitreous.	White.	White.
Compact, showing no trace of crystalline form or cleavage. — Little or no water in closed tube.	Amphibole (49). (Jade and Nephrite).	$(CaMgFe)SiO_3$.	Vitreous.	White to green.	White to greenish.
	Wollastonite (47).	$CaSiO_3$.	Vitreous.	White to reddish.	White.
Much water in closed tube.	Datolite (64).	$H_2Ca_2B_2Si_2O_{10}$.	Vitreous.	White to gray.	White.
	Calamine (69). Compare *Serpentine and Opal*.	$Zn_2SiO_4+H_2O$.	Vitreous.	White.	White.
Distinct tetragonal trisoctahedrons. — No water.	Leucite (57).	$K_2Al_2Si_4O_{12}$	Vitreous.	White to gray.	White.
Much water.	Analcite (72).	$Na_2Al_2Si_4O_{12}+2H_2O$.	Vitreous.	White.	White.
Very brittle to friable.	Apatite (36).	$3Ca_2P_2O_8+CaCl_2$.	Vitreous.	Green to brown, yellow, white, etc.	White.
Squarish, wedge-shaped or oblique crystals.	Titanite (65).	$CaTiSiO_5$.	Adamantine to resinous.	Brown, yellow, green, black.	White to gray.

H.	Tenacity.	G.	Form.	Cleavage.	Other Properties.	Confirmatory Chemical Tests.
5.	Tough.	2.8	V. Tabular or bladed to fibrous.	Distinct.	Translucent.	Fusible; gelatinizes with HCl.
5.	Brittle.	3.4	V. Sometimes bladed or fibrous.	Prismatic, perfect.	Translucent.	Fusible; insoluble.
5.	Brittle to flexible.	3.2	V. Usually bladed, fibrous or asbestiform.	Prismatic, perfect.	Translucent.	Fusible; insoluble.
5.	Brittle to flexible.	3.1	V. Usually bladed, fibrous or asbestiform.	Prismatic, perfect.	Translucent.	Fusible; insoluble.
5.	Brittle.	2.7	V. Acicular or fibrous and radiate.	Orthopinacoidal, perfect.	Translucent to opaque.	Fuses; yields water; gelatinizes with HCl.
5.25	Brittle.	2.4	IV. Acicular and radiate also massive.	Macropinacoidal, distinct.	Translucent. Pyroelectric.	Fuses with intumescence; yields water; gelatinizes with HCl.
5.25	Brittle.	2.2	IV. Acicular or fibrous and radiate.	Prismatic, perfect.	Transparent to translucent.	Fuses; yields water; gelatinizes with HCl.
5.	Tough.	3.	V. Compact.	None.	Translucent.	Fusible; insoluble.
5.	Tough.	2.8	V. Compact.	None.	Translucent.	Fuses; gelatinizes with HCl.
5.5	Brittle.	2.9	V. Globular and compact.	None.	Translucent to opaque.	Fuses with intumescence and colors flame bright green; much water; gelatinizes with HCl.
4.75	Brittle.	3.6	IV. Also botryoidal and massive.	Prismatic, perfect.	Translucent.	Infusible; yields water; green color with cobalt solution; gelatinizes with HCl.
5.5	Brittle.	2.5	I. Rarely massive.	None.	Translucent to opaque.	Infusible; blue color with cobalt solution.
5.25	Brittle.	2.25	I. Rarely massive.	Cubic, in traces.	Transparent to opaque.	Fuses; yields water; gelatinizes with HCl.
5.	Very brittle.	3.1	III. Prisms and massive or granular.	Basal, imperfect.	Transparent to opaque.	Fuses with difficulty; reaction for phosphorus with magnesium; soluble in acid.
5.	Brittle.	3.5	V. Crystals and cleavable massive.	Prismatic.	Translucent to opaque.	Fuses with intumescence.

V. Streak White or Gray.—Continued.
3. Hard.—Continued.
G. below 4.—Continued.

Analytical Key.			Species.	Composition.	Lustre.	Color.	Streak.
Crystalline or cleavable, and not belonging to the foregoing. Much water in closed tube. Crystalline or cleavable, and not belonging to the foregoing. No water in closed tube.		Octahedral cleavage. Cubic crystals.	Fluorite (18).	CaF_2.	Vitreous.	White, green, violet to blue, yellow, etc.	White.
		Gelatinizes with HCl.	Sodalite (57).	$3Na_2Al_2Si_3O_8 + 2NaCl$.	Vitreous to greasy.	Gray, blue, green, white, yellow, etc.	White.
			Nephelite (59).	$Na_2Al_2Si_2O_8$.	Vitreous to greasy.	White to gray or yellow.	White.
			Wollastonite (47).	$CaSiO_3$.	Vitreous.	White.	White.
		Fuses easily with intumescence.	Wernerite (58).	$CaAl_2Si_2O_8$.	Vitreous.	White, gray, reddish, etc.	White.
		Crystals and cleavage prismatic and rectangular.	Enstatite (47).	$MgSiO_3$.	Vitreous to pearly.	Grayish and greenish to brown.	White to gray.
			Pyroxene (47).	$(CaMgAl Fe)SiO_3$.	Vitreous.	Green to gray, brown, black, white.	White to greenish gray.
		Crystals and cleavage prismatic and oblique.	Amphibole (49).	$(CaMgAl \cdot Fe)SiO_3$.	Vitreous.	Green to white, gray, brown, black,	White to greenish gray.
		Effervesces and gelatinizes with HCl.	Cancrinite (59).	$Na_2Al_2Si_2O_8 + H_2O$ and CO_2.	Vitreous.	White, gray, yellow, red, blue, etc.	White.
		Gelatinizes with HCl. Fuses easily. G. below 3.	Gmelinite (73).	$Na_2Al_2Si_4O_{12} + 6H_2O$.	Vitreous.	White to greenish, reddish, etc.	White.
			Datolite (64).	$H_2Ca_2B_2Si_2O_{10}$.	Vitreous.	White to gray.	White.
		Gelatinizes with HCl. Infusible. G. above 3.	Calamine (69)	$Zn_2SiO_4 + H_2O$.	Vitreous.	White.	White.
		Fuses and colors flame violet	Apophyllite (70).	$4(H_2CaSi_2O_6) + KF$.	Vitreous to pearly.	White to yellowish or greenish.	White.
		Fuses with intumescence. G. below 2.5	Chabazite (73).	$K_2CaAl_2Si_5O_{15} + H_2O$.	Vitreous.	White to red.	White.
			Heulandite (74). Compare *Stilbite*.	$CaAl_2Si_6O_{16} + 5H_2O$.	Vitreous to pearly	White to red, gray, etc.	White.

H.	Tenacity.	G.	Form.	Cleavage.	Other Properties.	Confirmatory Chemical Tests.
4.	Brittle.	3.2	I. Cubes, also massive.	Octahedral, perfect.	Phosphorescent when heated.	Decrepitates; fuses and gives alkaline reaction; reaction for fluorine with $HKSO_4$.
5.5	Brittle.	2.3	I. Dodecahedrons and massive.	Dodecahedral, distinct.	Translucent.	Fuses with intumescence; gelatinizes with HCl.
5.5	Brittle.	2.6	III. Usually cleavable or massive.	Distinct.	Transparent to opaque.	Fusible; gelatinizes with HCl.
5.	Brittle to tough.	2.8	V. Cleavable or massive.	Distinct.	Translucent.	Fusible; gelatinizes with HCl.
5.	Brittle.	2.7	II. Also cleavable and massive.	Prismatic, distinct.	Transparent to opaque.	Fuses readily, with intumescence.
5.5	Brittle.	3.2	IV. Also massive and lamellar.	Prismatic, distinct.	Translucent to opaque.	Infusible and insoluble.
5.	Brittle.	3.4	V. Often massive and granular.	Prismatic, perfect.	Translucent to opaque.	Fusible; insoluble.
5.	Brittle.	3.2	V. Often bladed or massive.	Prismatic, perfect.	Translucent to opaque.	Fusible; insoluble.
5.	Brittle.	2.5	III. Usually massive.	Prismatic.	Translucent.	Fuses with intumescence; yields water; effervesces and gelatinizes with HCl.
4.5	Brittle.	2.1	III. Always in hexagonal crystals.	Prismatic, perfect.	Transparent to translucent.	Fuses with intumescence; much water; gelatinizes with HCl.
5.5	Brittle.	2.9	V. Glassy crystals.	Basal, distinct.	Translucent.	Fuses with intumescence, coloring flame bright green; much water; gelatinizes with HCl.
5.	Brittle.	3.6	IV. Also massive and granular.	Prismatic, perfect.	Transparent to opaque.	Infusible; yields water; green color with cobalt solution; gelatinizes with HCl.
5.	Brittle.	2.4	II. Square crystals, also massive.	Basal, perfect.	Transparent to opaque.	Fuses and colors flame violet; yields water; reacts for fluorine.
5.	Brittle.	2.1	III. Rhombohedrons.	Rhombohedral, distinct.	Transparent to translucent.	Fuses with intumescence; yields water.
4.	Brittle.	2.2	V. Also globular and granular.	Clinopinacoidal, perfect.	Transparent to opaque.	Fuses with intumescence; yields water.

Analytical Key.				Species.	Composition.	Lustre.	Color.	Streak.
V. Streak White or Gray.—Continued. / 4. Very Hard.	G. above 4.	Usually in distinct crystals.	G. 6.8	Cassiterite (26).	SnO_2.	Adamantine.	Brown to black and gray.	Gray to white.
				Rutile (26).	TiO_2.	Adamantine.	Brown to red.	Gray to brown.
			G. 4.2	Brookite (27).	TiO_2.	Adamantine.	Brown to yellowish and reddish.	White to gray.
				Garnet (53).	Complex silicate.	Vitreous.	All colors, often bright.	White to gray.
	G. below 4.	Crystallizes in slender prisms, or in columnar or bladed forms.	Yellowish and brownish green to dark brown.	Epidote (55).	$H_2Ca_4FeAl_2$ Si_6O_{26}.	Vitreous.	Yellowish green to dark brown.	Gray to white.
				Vesuvianite (54).	$(CaMg)_8$ $(AlFe)_4Si_7$ O_{28}.	Vitreous.	Brown to green.	White to gray.
				Tourmaline (62). Compare *Pyroxene* and *Amphibole*.	Complex silicate.	Vitreous.	Black, red, green, blue, etc.	White to gray.
			White, gray, light brown, blue, reddish, etc.	Diaspore (29).	Al_2O_3 $+H_2O$.	Pearly to vitreous.	White to gray and brown.	White.
				Cyanite (63).	Al_2SiO_5.	Vitreous to pearly.	Blue to white, green and gray.	White.
				Fibrolite (63).	Al_2SiO_5.	Vitreous.	Brown to gray, white, green, etc.	White.
				Zoisite (55). Compare *Pyroxene* and *Amphibole*.	$H_2Ca_4Al_6$ Si_6O_{26}.	Vitreous to pearly.	Gray, brown, green, red.	White.
		Compac, showing no race of crystalline form or cleavage.	Not easily or distinctly scratched by vitreous quartz.	Quartz (Chalcedony, Carnelian, Chrysoprase, etc.) (33).	SiO_2.	Waxy.	White gray brown, red, green, etc.	White.
				Quartz (Agate, Onyx, etc.) (33).	SiO_2.	Waxy.	Same, but banded, clouded, or dendritic.	White.
				Quartz (Jasper, etc.) (33).	SiO_2.	Waxy to dull.	Red, brown, yellow, green, black, etc.	White to gray.
				Quartz (Flint, Chert, etc.) (33). Compare *Vitreous* and *Milky Quartz*.	SiO_2.	Waxy to dull.	Gray, brown, black.	White to gray.

H.	Tenacity.	G.	Form.	Cleavage.	Other Properties.	Confirmatory Chemical Tests.
6.5	Brittle.	6.8	II. Square prisms, botryoidal and massive.	Indistinct.	Translucent to opaque.	Infusible and insoluble; tin with soda on charcoal.
6.5	Brittle.	4.2	II. Usually in twins.	Prismatic, distinct.	Translucent to opaque.	Infusible and insoluble.
6.	Brittle.	4.2	IV. Small square crystals.	Imperfect.	Translucent.	Infusible and insoluble.
6.5	Brittle.	4.2	I. Crystals, rarely massive.	None.	Transparent to opaque.	Fusible, usually to a magnetic globule; insoluble.
6.5	Brittle.	3.4	V. Crystals usually six-sided.	Orthopinacoidal, perfect.	Transparent to opaque.	Fuses with intumescence to a magnetic mass.
6.5	Brittle.	3.4	II. Square prisms, often divergent.	Indistinct.	Translucent.	Fuses with intumescence.
7.	Brittle.	3.1	III. Rarely massive.	None.	Transparent to opaque.	Mostly infusible; insoluble.
6.5	Brittle.	3.4	IV. Usually bladed or foliated.	Brachypinacoidal, perfect.	Translucent.	Infusible; decrepitates strongly; blue color with cobalt solution; insoluble.
7.	Brittle.	3.6	VI. Coarsely bladed.	Pinacoidal, distinct.	Translucent.	Infusible and insoluble; blue color with cobalt solution.
6.5	Brittle.	3.2	V. Usually bladed or fibrous.	Orthopinacoidal, perfect.	Translucent.	Like cyanite.
6.5	Brittle.	3.3	IV. Slender and deeply striated.	Brachypinacoidal, perfect.	Translucent.	Fuses with intumescence.
7.	Brittle.	2.6	III. Botryoidal, stalactitic, etc.	Conchoidal.	Translucent.	Infusible; insoluble; dissolves with effervescence in soda on platinum wire.
7.	Brittle.	2.6	III. Botryoidal, geoditic, etc.	Conchoidal.	Translucent.	Like chalcedony.
7.	Brittle.	2.6	III. Compact, banded, etc.	Conchoidal.	Opaque.	Like chalcedony.
7.	Brittle.	2.6	III. Usually in irregular nodules.	Conchoidal.	Translucent to opaque.	Like chalcedony.

Analytical Key.					Species.	Composition.	Lustre.	Color.	Streak.
V. Streak White or Gray.—Continued.	4. Very Hard.—Continued.	G. below 4.—Continued.	Compact, showing no trace of crystalline form or cleavage.—Continued.	Bright blue.	Turquois (38).	$Al_2P_2O_{11}$ $+5H_2O$.	Waxy to dull.	Sky-blue to apple-green.	White to greenish.
					Lapis Lazuli (57). Compare *Lazulite*.	$(CaNaAl)_2$ SiO_4.	Vitreous.	Blue.	Blue.
				Green.	Epidote (55).	$H_2Ca_4FeAl_2$ Si_6O_{26}.	Vitreous.	Yellowish green.	Gray.
					Amphibole (49). (Jade and Nephrite). Compare *Chrysolite*.	$(CaMgFe)$ SiO_3.	Vitreous.	Deep green to greenish white.	Greenish to white.
				Water in closed tube. G. 2.1	Opal (34).	SiO_2+H_2O.	Vitreous to pearly.	Various colors.	White.
				Red to brown. Weathers black. G. 3.5	Rhodonite (48).	$MnSiO_3$.	Vitreous.	Red and brown to gray.	White.
				Yellow to gray, brown, etc. G. above 3.	Chondrodite (61).	$Mg_3Si_3O_{14}$.	Vitreous to resinous.	Yellow and brown to red and green.	White to gray.
					Zoisite (55).	$H_2Ca_4Al_6$ Si_6O_{26}.	Vitreous to dull.	Gray to brown and green.	White to gray.
				Many colors. Weathers white. G. below 3.	Feldspar family (60).	See next table.	Vitreous to dull.	Various colors.	White to gray.
			Crystalline or cleavable.	Crystals distinctly isometric.	Leucite (57).	$K_2Al_2Si_4O_{12}$.	Vitreous.	White to gray.	White.
					Garnet (53).	Complex silicate.	Vitreous.	All colors, often bright.	White to gray.
				Crystals hexagonal prisms and pyramids.	Quartz (Amethyst, Smoky, Ferruginous, etc.) (33).	SiO_2.	Vitreous.	Colorless, purple, yellow, brown, etc.	White.
				Crystals distinctly prismatic and approximately hexagonal or triangular.	Tourmaline (62).	Complex silicate.	Vitreous.	Black, red, green, blue, etc.	White to gray.
					Epidote (55).	$H_2Ca_4FeAl_2$ SiO_6O_{26}.	Vitreous.	Yellowish green to dark brown.	Gray to white.
					Staurolite (66).	$MgFe_2Al_{12}$ Si_6O_{34}.	Vitreous to resinous.	Dark brown to black.	Gray to white.

H.	Tenacity.	G.	Form.	Cleavage.	Other Properties.	Confirmatory Chemical Tests.
6.	Brittle.	2.7	Amorphous, incrusting.	None.	Opaque.	Infusible; but yields water and turns brown; soluble in HCl.
6.5	Brittle.	2.4	I. Usually compact.	None.	Opaque.	Fusible; gelatinizes with HCl.
6.5	Brittle.	3.4	V. Compact to finely granular.	None.	Opaque.	Fuses with intumescence to a magnetic mass.
6.	Tough.	3.2	V. Compact.	None.	Translucent.	Fusible; insoluble.
6.	Brittle.	2.1	Amorphous.	None.	Transparent to translucent.	Infusible; water in closed tube.
6.5	Tough.	3.5	VI. Compact or massive.	Indistinct.	Translucent to opaque.	Fusible; amethystine bead with borax.
6.5	Brittle.	3.2	V. Usually in rounded grains.	None.	Translucent to opaque.	Infusible; gelatinizes with HCl.
6.5	Brittle.	3.3	IV. Compact to finely crystalline.	None.	Translucent to opaque.	Swells up and fuses.
6.5	Brittle.	2.7	V. and VI. Compact and porphyritic.	None.	Opaque.	Fuses with difficulty; insoluble; weathers white.
6.	Brittle.	2.5	I. Crystals, rarely massive.	None.	Translucent to opaque.	Infusible; blue color with cobalt solution; insoluble.
6.5	Brittle.	3.2	I. Crystals, rarely masive.	None.	Transparent to opaque.	Fusible, usually to a magnetic globule; insoluble.
7.	Brittle.	2.6	III. Prisms and pyramids.	None.	Transparent to opaque.	Infusible; insoluble; dissolves with effervescence in soda on platinum wire.
7.	Brittle.	3.1	III. Usually triangular prisms.	None.	Transparent to opaque.	Mostly infusible; insoluble.
6.5	Brittle.	3.4	V. Crystals usually six-sided.	Orthopinacoidal, perfect.	Transparent to opaque.	Fuses with intumescence to a black, magnetic mass.
7.	Brittle.	3.6	IV. Usually in cruciform twins.	Imperfect.	Translucent to opaque.	Infusible and insoluble.

V. Streak White or Gray.—Continued. / *4. Very Hard.—Continued.* / *G. below 4.—Continued.* / Crystalline or cleavable, and not belonging to the foregoing.

Analytical Key.	Species.	Composition.	Lustre.	Color.	Streak.
Crystals distinctly prismatic and approximately square or octagonal.	**Wernerite** (58).	$CaAl_2Si_2O_8$.	Vitreous.	White, gray, reddish, etc.	White.
	Vesuvianite (54).	$(CaMg)_8(AlFe)_4Si_7O_{28}$.	Vitreous.	Brown to green.	White to gray.
	Andalusite (63).	Al_2SiO_5.	Vitreous to dull.	White, gray, red, green, brown.	White to gray.
Crystals distinctly prismatic and not belonging to the foregoing.	**Zoisite** (55).	$H_2Ca_4Al_6Si_6O_{26}$.	Vitreous.	Gray, to brown, green to red.	White.
	Spodumene (48).	$Li_6Al_8Si_{15}O_{45}$.	Vitreous to pearly.	White to gray and green.	White.
Blue.	**Lazulite** (38).	$MgAlP_2O_9+H_2O$.	Vitreous.	Blue.	White.
No distinct cleavage. Green to greenish gray.	**Chrysolite** (52).	$(MgFe)_2SiO_4$.	Vitreous.	Green.	White.
	Prehnite (69).	$H_2CaAl_2Si_3O_{12}$.	Vitreous.	Green to gray.	White.
No distinct cleavage. Red, yellow or brown.	**Rhodonite** (48).	$MnSiO_3$.	Vitreous.	Red and brown to gray.	White.
	Quartz (Amethyst, Rose, Smoky and Ferruginous) (33).	SiO_2.	Vitreous.	Purple, red, yellow, brown, etc.	White.
	Chondrodite (61).	$Mg_8Si_3O_{14}$.	Vitreous to resinous.	Yellow, red, brown, to greenish.	White.
No cleavage. White or colorless.	**Quartz** (Vitreous, Milky, etc.) (33).	SiO_2.	Vitreous.	All colors and colorless.	White.
Cleavable obliquely. G. above 3.	**Spodumene** (48).	$Li_6Al_8Si_{15}O_{45}$.	Vitreous to pearly.	White to gray and green.	White.
Cleavable at right angles. G. below 3.	**Feldspars** (60). **Orthoclase** and **Albite**.	$K_2Al_2Si_6O_{16}$. $Na_2Al_2Si_6O_{16}$.	Vitreous to pearly.	White to gray, red, green, etc.	White.
	Feldspars (60). **Oligoclase** and **Labradorite**. Compare *Nephelite* and *Wernerite*.	$(Na_2Ca)Al_2Si_5O_{14}$. $(Na_2Ca)Al_2Si_3O_{10}$.	Vitreous to pearly.	White, gray to greenish and reddish.	White.

H.	Tenacity.	G.	Form.	Cleavage.	Other Properties.	Confirmatory Chemical Tests.
6.	Brittle.	2.7	II. Also cleavable and massive.	Prismatic, distinct.	Transparent to opaque.	Fuses readily with intumescence.
6.5	Brittle.	3.4	II. Often irregular and divergent.	Indistinct.	Translucent.	Fuses with intumescence.
6.	Brittle.	3.2	IV. Usually black square or cross on section.	Imperfect.	Transparent to opaque.	Infusible and insoluble.
6.5	Brittle.	3.3	IV. Slender prisms deeply striated.	Brachypinacoidal, perfect.	Translucent.	Fuses with intumescence.
6.75	Brittle.	3.2	V. Large broad crystals.	Prismatic, perfect.	Transparent to opaque.	Fuses with reddish flame and intumescence.
6.	Brittle.	3.	V. Octahedrons, also massive.	Imperfect.	Opaque.	Whitens and yields water; blue color is restored by cobalt solution.
6.5	Brittle.	3.4	IV. Usually in grains or granular.	Conchoidal.	Translucent.	Infusible; gelatinizes with HCl.
6.5	Brittle.	2.9	IV. Often with crested surface.	Basal, indistinct.	Translucent.	Fusible with intumescence; yields water.
6.5	Tough.	3.5	Compact to finely crystalline.	None.	Translucent to opaque.	Fusible; amethystine bead with borax.
7.	Brittle.	2.6	III. Prisms, pyramids and massive.	Conchoidal.	Transparent to opaque.	Infusible; insoluble; dissolves with effervescence in soda on platinum wire.
6.5	Brittle.	3.2	V. Usually in rounded grains.	None.	Translucent to opaque.	Infusible; gelatinizes with HCl.
7.	Brittle.	2.6	III. Prisms, pyramids and massive.	Conchoidal.	Transparent to opaque.	Infusible; insoluble; dissolves with effervescence in soda on platinum wire.
6.75	Brittle.	3.2	V. Prisms and cleavable masses.	Prismatic, perfect.	Translucent to opaque.	Fuses with reddish flame and intumescence.
6.5	Brittle.	2.6	V. and VI. Often massive.	Basal and clinopinacoidal, perfect.	Transparent to translucent.	Fusible with difficulty; insoluble.
6.5	Brittle.	2.7	VI. Usually massive.	Like orthoclase.	Transparent to translucent.	Fusible with difficulty; insoluble.

V. Streak White or Gray.—Continued.

5. Adamantine.

H. not above 8, do not scratch beryl distinctly.

H. above 8, scratch beryl.

Analytical Key.			Species.	Composition.	Lustre.	Color.	Streak.
H. 10.		Isometric.	Diamond (5).	C.	Adamantine, brilliant.	Colorless to yellowish, reddish, etc.	White.
H. 9.		Hexagonal.	Corundum (25). Compare *Emery.*	Al_2O_3.	Vitreous.	Gray, brown, red, yellow, blue, black, etc.	White.
H. 8.5		Orthorhombic.	Chrysoberyl (24).	$BeAl_2O_4$.	Vitreous.	Green.	White.
	Octahedrons.	Black to red.	Spinel (23).	$MgAl_2O_4$.	Vitreous.	Black, red, blue, green, yellow, etc.	White.
	Dodecahedrons and tetrag. trisoct.	Chiefly red, yellow or brown.	Garnet (53).	Complex silicate.	Vitreous.	All colors, often bright.	White.
	Square prisms and pyramids.	Adamantine usually bright.	Zircon (26).	$ZrSiO_4$.	Adamantine.	Gray, yellowish, brownish, etc.	White.
		Vitreous, usually weak.	Andalusite (63).	Al_2SiO_5.	Vitreous to dull.	White, gray, red, green, brown.	White.
	Triangular prisms and rhombohedrons.	Chiefly black.	Tourmaline (62).	Complex silicate.	Vitreous.	Black, red, green, blue, etc.	White to gray.
	Rhombic prisms.	Perfect basal cleavage.	Topaz (64).	$Al_2SiO_4F_2$.	Vitreous.	Yellow, white, green, blue, red, etc.	White.
		No basal cleavage.	Staurolite (66).	$MgFe_2Al_{12} Si_6O_{34}$.	Vitreous to resinous.	Dark brown to black.	White to gray.
	Hexagonal prisms or Pyramids.	Green.	Beryl (50). Compare *Epidote.*	$Be_3Al_2Si_6 O_{18}$.	Vitreous.	Green to yellow and white.	White.
		Various colors.	Quartz (Amethyst, Smoky, Ferruginous, etc.)(33).	SiO_2.	Vitreous.	Colorless, purple, yellow, brown, etc.	White.
	Distinctly bladed or cleavable.	Blue to gray, etc.	Cyanite (63). Compare *Fibrolite, Diaspore, Topaz.*	Al_2SiO_5.	Vitreous to pearly.	Blue to white, green and gray.	White.
	No distinct cleavage.	Green.	Chrysolite (52).	$(MgFe)_2 SiO_4$.	Vitreous.	Green.	White.
		Brown to black.	Cassiterite (26).	SnO_2.	Adamantine.	Brown to black, etc.	White to gray.

H.	Tenacity.	G.	Form.	Cleavage.	Other Properties.	Confirmatory Chemical Tests.
10.	Brittle.	3.5	I. Usually with curved faces	Octahedral, perfect.	Transparent.	Burns at a high temperature; insoluble.
9.	Brittle to tough.	4.	III. Also massive to finely granular.	Basal and rhombohedral.	Transparent to opaque.	Infusible and insoluble.
8.5	Brittle.	3.7	IV. Usually in hexagonal twins.	Imperfect.	Transparent to translucent.	Infusible and insoluble.
8.	Brittle.	3.8	I. Octahedrons and water-worn grains.	Octahedral.	Transparent to opaque.	Infusible and insoluble.
7.5	Brittle.	3.2 to 4.3	I. Distinct crystals, rarely granular.	Imperfect.	Transparent to opaque.	Fusible, usually to a magnetic globule; insoluble.
7.5	Brittle.	4.4	II. Rarely irregular grains.	Imperfect.	Transparent to opaque.	Infusible and insoluble.
7.5	Brittle.	3.2	IV. Usually a black square or cross on section.	Imperfect.	Transparent to opaque.	Infusible and insoluble.
.7.5	Brittle.	3.1	III. Rarely massive.	None.	Transparent to opaque.	Mostly infusible; insoluble.
8.	Brittle.	3.5	IV. Rarely massive or granular.	Basal, perfect.	Transparent to opaque.	Infusible and insoluble; blue color with cobalt solution; fluorine reaction.
7.25	Brittle.	3.6	IV. Usually in cruciform twins.	Imperfect.	Translucent to opaque.	Infusible and insoluble.
8.	Brittle.	2.7	III. Rarely massive.	Imperfect.	Transparent to translucent.	Infusible and insoluble.
7.	Brittle.	2.6	III. Prisms and pyramids.	None.	Transparent to opaque.	Infusible; insoluble; dissolves with effervescence in soda on platinum wire.
7.25	Brittle.	3.6	VI. Coarsely bladed.	Pinacoidal, distinct.	Translucent.	Infusible; insoluble; blue color with cobalt solution.
7.	Brittle.	3.4	IV. Usually in grains or granular.	Conchoidal.	Translucent.	Infusible; gelatinizes with HCl.
7.	Brittle.	6.8	II. Square prisms, botryoidal and massive.	Indistinct.	Translucent to opaque.	Infusible; insoluble; tin with soda on charcoal.

HOW TO USE THE TABLES.

The plan of these Tables, and the method of using them in determining minerals, have been explained incidentally in the Introduction; but a more connected statement will probably be useful to many students. Minerals are divided at the outset into two great classes, *metallic* and *non-metallic*. Each class comprises five subclasses, the metallic subclasses being distinguished according to the *colors* of the species, and the non-metallic subclasses according to the *streaks*, or colors of the powdered minerals. The species in each subclass are further distinguished, according to hardness, as *very soft, soft, hard, very hard* and, in the fifth non-metallic subclass, *adamantine*. We thus by (1) the lustre, (2) the color or streak and (3) the hardness, of minerals divide them into forty-one groups, as shown in the general classification on page 25. This synopsis of the tables also obviates the necessity of continually turning the pages, as it enables us to turn at once to the page on which the mineral in hand is described.

In the analytical key on the left margin of each table these groups are subdivided in accordance with other and various physical properties, until we come to the individual species, or groups of two or three species only, in the next column to the right. While in the succeeding columns we find, besides the composition, a concise physical description of each species, by·which it may be more carefully distinguished from those most closely resembling it, and the identification verified; and, finally, the broad column on the right margin contains a brief statement of the chemical tests, which are our last resort, to be tried only when the determination is not otherwise satisfactory.

To make the method of using the tables still clearer, a single example is appended, using a fragment of galenite. Turning to the general classification on page 25 , we observe first that it is *metallic*, and take the left column; next that it is not red, brown, yellow or ·black, but *gray*, belonging in the fourth subclass; and then, trying the hardness with steel, find that it cuts easily and is *soft*; and turn to page 42 for the detailed analysis of this group. Here the six species in this group are divided by the streak into three subgroups, and on testing the specimen it is found to belong to the second, having a *dark gray streak*. This subgroup includes three species—stibnite, galenite and chalcocite; and on glancing over the physical descriptions to the right it is seen that, while they agree closely in certain properties, they are easily distinguished by others, especially by color, density, form and cleavage. If, however, the specimen is of an impure or doubtful character, fusion on charcoal will show that it gives reactions for lead, but not for copper and antimony.

In all cases where the meaning of the Tables is not clear, the student should refer to the section of the Introduction where the property or test is explained. It may be fairly said that the golden rule in determinative mineralogy is to *follow the order of the tables*, and not skip about, or guess at the names or relations of minerals.

BOSTON SOCIETY OF NATURAL HISTORY.

TEACHERS' SCHOOL OF SCIENCE.

MINERALOGICAL AND GEOLOGICAL

SPECIMENS AND COLLECTIONS.

PREPARED FOR THE USE OF TEACHERS AND STUDENTS.

It is now generally recognized by the best teachers that satisfactory results in the study of natural science can only be attained by the liberal use of specimens; and unquestionably the difficulty and expense of obtaining suitable material has been a great obstacle to the introduction of the natural or scientific method. The vital importance of this matter to the interests of sound education is fully appreciated by Prof. Alpheus Hyatt, of the Teachers' School of Science; and with his aid and encouragement the collections described below have been prepared for the express purpose of supplying schools, teachers and students with carefully selected educational material at the minimum cost.

ELEMENTARY MINERALOGY.

These collections are designed to illustrate Science Guide No. XIII. ("First Lessons in Minerals," by Ellen H. Richards).

Collection No. 1 includes the twenty principal elements and minerals of which it is important to have one specimen for each pupil.

1 large specimen of each kind, 20 in all, labelled.	.	$ .50.
5 smaller specimens of each kind, 100 in all,	. .	. 1.25.

Collection No. 2 includes collection No. 1 and ten additional varieties of which it is desirable to have at least one specimen for every two or three pupils.

1 large specimen of each kind, 10 in all, labelled,	.	$.30.
5 smaller specimens of each kind, 50 in all,	. . .	.75.

MINERALOGY.

The mineralogical collections may be used advantageously with Prof. Dana's Manual of Mineralogy, or Text-book of Mineralogy, or any standard text-book.

Scale of Hardness.—Price: Cabinet size, $1.00; Student size, 50 cents.

Lustre Series.— Six specimens illustrating the principal kinds of lustre.

Price: Cabinet size, $1.00.

Cleavage Series.—Ten specimens illustrating the principal kinds of cleavage in the different systems of crystallization. Price: Cabinet size, 1.50.

Descriptive Mineralogy Series.—

	50 specimens.	100 specimens.	150 specimens.
Cabinet size, ...	$6.00.	$15.00	$30.00.
Student size, ...	. 2.00.	5.00	10.00.

Miniature Mineral Collections in neat, covered paste-board boxes or trays.

These minerals are selected from the choicest materials to be had, and with special reference to their purity, beauty and crystalline form. Each specimen is numbered to correspond with the numbers in a printed list in the cover of the box, and also with a descriptive hand-book that accompanies each collection.

No. 1. Twenty-five specimens, prepaid by mail,	. .	$1.00.
No. 2. Fifty specimens, " " "	. ,	2.00.

DETERMINATIVE MINERALOGY.

Many teachers and students using the first edition of the "Tables for the Determination of Common Minerals" have applied for specimens suitable for practice in the determination of species. To meet this demand in the future, small, but pure and typical specimens of most of the minerals included in the scope of the "Tables" will be furnished at an average price of 3 cents each, with a large discount for duplicates.

Pure minerals suitable for blowpipe analysis and chemical experiments are sold by weight, the prices ranging from 5 cts. to 50 cts. per pound.

LITHOLOGY.

The lithological collections were originally prepared to illustrate Science Guide No. XII. (Common Minerals and Rocks, by W. O. Crosby;) but they have been recently considerably extended, and the larger ones, especially, may be advantageously used in connection with more advanced text-books.

The prices of these collections are as follows:

	50 spec's.	80 spec's.	125 spec's.	150 spec's.
Cabinet size with printed labels,	$2.00.	$4.00.	$8.00.	$10.00.
Student size,	1.25.	2.50.	5.00.	6.00

The specimens in the Student collections are not labelled but are numbered to correspond with the printed catalogue.

ECONOMIC MINERALOGY AND LITHOLOGY.

Many teachers desire to give special prominence to those minerals and rocks having important uses in the arts. To meet this need, the following collections have been arranged (1). **Ores.**—This collection includes 30 typical specimens of the most important ores. It embraces ores of gold, silver, mercury, copper, lead, zinc, tin, iron, etc. Price: Cabinet size, $5.00.

Economic Minerals other than Ores.—This collection includes 45 specimens of minerals having important uses in the arts, but from which no metal is obtained, such as sulphur, graphite, corundum, gypsum, apatite, barite, halite, asbestus, etc. Price: Cabinet size, $1.50. Collections 1 and 2 will be sold together for $9.00; 1, 2 and 3 for $10.00.

STRUCTURAL GEOLOGY OR PETROLOGY.

This collection consists of 30 specimens illustrating nearly all the most important kinds of structures occuring in rocks, as follows: Stratification, Ripple-marks, Rain-prints Mud-cracks, Fossils, Veins, Dikes, Stalactites, Joints, Cleavage, Faults, Folds and Contortions, Concretions, Glacial Striæ, etc. Price: Cabinet size $8.00.

HISTORICAL GEOLOGY.

Stratigraphic Collection.—This includes 100 specimens of the characteristic rocks of the various geological formations from the Laurentian to the Tertiary. Price: Cabinet size, $8.00. **Paleontological Collection.**—This embraces 50 species (about 100 specimens) of fossils, selected from the characteristic forms of the different formations. Price: Cabinet size, $8.00.

APPARATUS.

The Apparatus required in using these determinative tables will be furnished as follows:— Brass Blowpipe, .20; Alcohol lamp, .50, Bunsen burner, 75; Steel forceps, 10; Platinum wire 10; Glass tubes, open, .05 per dozen; closed, .10 per dozen; Hammer, anvil and ring, .40; File, .10; Magnet, .15; Lens, .75; Test tubes, .05 each.

All orders should be addressed to

GEORGE H. BARTON,
Boston Society of Natural History,

Cor. Berkeley and Boylston Sts.   BOSTON, Mass.

CPSIA information can be obtained
at www.ICGtesting.com
Printed in the USA
BVHW041528291118
534322BV00009B/97/P